TJ
211
•N16
1987

N—Nagy, Francis.

Engineering
   foundations of
   robotics

$39.33

DATE		

ENGINEERING FOUNDATIONS OF ROBOTICS

ENGINEERING FOUNDATIONS OF ROBOTICS

Francis N-Nagy
PhD

*Department of Electronic
and Electrical Engineering
University of Salford
Salford, United Kingdom*

Andras Siegler
Dr Eng

*Computer and Automation Institute
Department of Robotics and Computer Vision
Hungarian Academy of Sciences
Budapest, Hungary*

Prentice-Hall International

Englewood Cliffs, NJ   London
Mexico   New Delhi   Rio de Janeiro
Singapore   Sydney   Tokyo   Toronto

Library of Congress Cataloging in Publication Data

N-Nagy, Francis.
  Engineering foundations of robotics.

  Bibliography: p.
  Includes index.
  1. Robotics.   I. Siegler, Andras, 1952–
II. Title.
TJ211.N16   1986   629.8′92   86-5073
  ISBN 0-13-278805-5

British Library Cataloguing in Publication Data

N-Nagy, Francis
  Engineering foundations of robotics.
  1. Robots, Industrial
  I. Title   II. Siegler, Andras
  629.8′92   TS191.8

  ISBN 0-13-278805-5
  ISBN 0-13-278797-0 Pbk

©1987 Prentice-Hall International (UK) Ltd

Prentice-Hall Inc., *Englewood Cliffs, New Jersey*
Prentice-Hall International (UK) Ltd, *London*
Prentice-Hall of Australia Pty Ltd, *Sydney*
Prentice-Hall Canada Inc., *Toronto*
Prentice-Hall Hispanoamericana S.A., *Mexico*
Prentice-Hall of India Private Ltd, *New Delhi*
Prentice-Hall of Japan Inc., *Tokyo*
Prentice-Hall of Southeast Asia Pte Ltd, *Singapore*
Editora Prentice-Hall do Brasil Ltda, *Rio de Janeiro*

Printed and bound in Great Britain for
Prentice-Hall International (UK) Ltd,
66 Wood Lane End, Hemel Hempstead,
Hertfordshire, HP2 4RG
at the University Press, Cambridge

1 2 3 4 5   90 89 88 87 86

0-13-278805-5
0-13-278797-0 PBK

CONTENTS

FOREWORD

This book is a welcome addition to the increasing volume of literature on robotics. Although the subject is still in its relative infancy, there has been an almost explosive growth in the application of robots in recent years ranging from relatively simple pick-and-place devices to complex multi-degree-of-freedom machines with computer assisted programming of their functions. This rapid growth in the industrially installed base of robots and the accompanying development of the supporting technologies has not been matched by a full range of good and comprehensive textbooks covering this vital area of automation.

I have known one of the authors, Frank N-Nagy, for over twenty years, and he can only be described as an out and out enthusiast for robotics and its further development. Frank and Andras, his co-author, have assembled a comprehensive book which takes the reader through an overview of robotics and a wide ranging series of chapters which present the underlying mathematical techniques and principles relating to the kinematic models of robot manipulators.

The need for new designs of robots is of paramount importance since today's robots are quite primitive in terms of onboard intelligence, mechanism design, stability, controllability, and speed of response. Generally, they are still far too slow, particularly for applications in the rapidly expanding area of assembly automation, and they can be quite inaccurate. They often have poor trajectory-following performance with inadequate control of both velocity and acceleration. In many cases, power to weight ratios are inadequate, and substantial improvements in performance can certainly be achieved through the use of better servomotors and of lighter but stiffer structures.

Thus, this book is being published at a particularly opportune time, and it is likely to become a standard reference book for robot designers. A feature of the text is the incorporation of twelve *Illustrative examples* which are each complete in themselves and which should be of considerable assistance to designers embarking on the creation of new kinematic designs and their controllers.

Roger Crossley

Professor of Manufacturing Systems Engineering
Department of Aeronautical and Mechanical Engineering
University of Salford, UK

and

Director, AMTeC
The Advanced Manufacturing Technology Centre, UK

PREFACE

Examples are more useful than rules.
Isaac Newton

Nothing is as practical as a good theory.
Ludvig Boltzman

These great principles of Newton and Boltzman have governed the authors throughout the preparation of this book. Both authors graduated as engineers, and have since spent a considerable time in industry and professional institutions. They appreciate that engineers in industry have limited time for digging out reference material in libraries or patent offices and that students and research scientists have limited financial support to attend conferences and seminars to familiarize themselves with advances in new engineering fields. They just cannot afford time for lengthy exploration. For them and other workers in the field of robotics the authors have endeavored to bring the most needed material together and to bridge the gaps in robotics.

Both authors have been directly involved in industrial robot development work. One of the authors (F. N-Nagy) initiated and provided the technical leadership for the design and development of the first British lightweight anthropomorphic industrial robot, commissioned for Syke Instrumentation Ltd, which is now on the market. The other author (A. Siegler) has sound expertise in robot programming and in CAD/CAM systems design including robot application. The authors' experience is reflected in the engineering content throughout the book.

This book presents the fundamental mathematical tools used in the design and control of robot position and motion, and for devising robot language commands for synthesizing tasks. Objectives, like control of robot manipulators by both direct and inverse homogeneous transformation methods, the task of moving objects in world and tool coordinate systems, various ways of teaching robots, integration of teaching facilities into task programs, writing and executing structured robot programs, control, etc. are developed in case studies presented in the form of Illustrative examples.

The material is presented in seven chapters. Chapter 1 presents a review of robotics, gives the definitions of concepts in robot engineering, and discusses possible trends in standardization including MAP/TOP protocols. It is also a supportive text to the rest of the book. Chapter 2 deals with homogeneous coordinate transformations. The matrix description of objects in three dimensions is also dealt with in this chapter. Chapter 3 discusses Euler angles of orientation, and revolute, cylindrical and polar coordinate systems for describing robot motion. In Chapter 4 object manipulation in the robot workspace and design methods of motion sequence are developed. The use of visual information and the design of a robot workstation equipped with a vision system are introduced to show the implications of trajectory planning. Chapter 5 presents the calculation of tool's paths in terms of joint variables. Chapter 6 deals with the derivation of motion in the tool coordinate systems by applying the inverse homogenous transformation. A useful method for designing the effective working area of the robot manipulator is developed. Chapter 7 shows the derivation of basic servo theories tailored for robot use. It includes the interpretation of figures of merit in robot servo systems. Feedforward and feedback compensations are discussed.

The reading and study of this book should cause no difficulty to anyone with a basic theoretical and practical engineering background. The authors have aimed to build on such a background in explaining and deriving the mathematical description of robot operations. Nonetheless, wherever complementary theoretical support is felt to be necessary, further clarification is given in the appendices.

November 1986   F. N-Nagy
                                                                A. Siegler

ACKNOWLEDGEMENTS

F. N-Nagy wishes to thank the Directors of Syke Instrumentation Company for commissioning the feasibility study on the design of a lightweight industrial robot. Special thanks are due to Sue Clarke, Managing Director of the Company, who procured the financial support which enabled the essential research and development for the production of the first British designed and manufactured industrial anthropomorphic robot. This great opportunity inspired F. N-Nagy to write this book.

A. Siegler expresses his thanks to the Computer and Automation Institute, Academy of Sciences, Hungary, for enabling him to partake in writing this book.

We would like to thank Professor Roger Crossley for his advice and comments on the manuscript, and Professor John Gray for his support in establishing the Robotics and Mechatronics Laboratory in the Electronic and Electrical Engineering Department at Salford University where the material presented here was partly laboratory-tested.

The authors wish to acknowledge the manufacturers, scientists and students who have been helpful in the preparation of this text. Special thanks must go to colleagues Dr Jeno Takacs (University of Oxford), John Haslam (Technical College of Stockport), Constantine Mousakis (University of Salford), and to postgraduate students William Edmonds, John Kennedy, Richard Udo, John Brazendale and many others on the Electronic Control Engineering course at the University of Salford where most of the material has at some time been included in the MSc course curricula.

NOTATION

$*$ = matrix by matrix multiplication
$\circ$ = scalar multiplication
χ = vector multiplication
$\cos q_i = ci$
$\sin q_i = si$

1

INTRODUCTION: OVERVIEW

1.1 INTRODUCTION

The authors' main objective is to introduce the wide-ranging topics of robotics – a subject which is not just a collection of selected segments of engineering science, but a coherent presentation of the fundamentals of a developing theory within the new fields of automation.

Because of the rapid changes taking place in robot technology, there is a need to devise theoretical tools and practical means of enabling industry to progress from the current state of automation into robotization. The aim here is to give a general explanation of how a robot operates and to provide easily applicable theories that can form the basis of usable design methods for robot operations. The theories are expounded in such a way that they require neither a deep involvement in complex mathematics, not an expert knowledge of how to implement mathematics in robotics. Various newly devised design methods make the application of the theories even easier by the use of Illustrative examples.

1.2 WHAT IS AND WHAT ISN'T A ROBOT

In this section we provide some description of robots' functions and deal with their desirable features. To understand the functional concept, the following questions have to be answered: what is and what isn't a robot? How is a robot constructed? How does it operate? Or, quite simply: What is the definition of a robot?

Despite the fact that a wide spectrum of definitions exists, few manufacturers or users would agree on any one. In fact, none has been accepted as standard. Since there is no standard either for defining or for describing the functions of a robot, it would be helpful if we first consider some of the attempts to provide one.

The British Robot Association (BRA) has defined the industrial robot as

> a reprogrammable device with a minimum of four degrees of freedom designed to both manipulate and transport parts, tools or specialized manufacturing implements through variable programmed motions for the performance of the specific manufacturing task.

One of the important qualifications of the British definition is the four degrees of freedom. This is important for their selective distinction.

The Robotics Institute of America (RIA) defines the robot as a reprogrammable, multi-functional manipulator designed to move material, parts, tools, or specialized devices through variable programmed motions for the performance of a variety of tasks.

This is a widely accepted definition of an industrial robot. The emphasis is on the programmable facilities by which the robot can perform simple tasks without human assistance.

Japan Industrial Robot Association (JIRA) and the Japanese Industrial Standards Committee [1.1] in the 'Glossary of terms for industrial robots' (*JIS* B 0134-1979) defines the robot at various levels as

manipulator: a machine which has functions similar to those of the human upper limbs, and moves the object spatially, from one location to the other [No. 1101]* . . . playback robot: a manipulator which is able to perform an operation by reading off the memorized information for an operating sequence, including positions and the like, which it learned by being taken manually through the routine beforehand [No. 1106] . . . intelligent robot: a robot which can determine its own behaviour/conduct through its functions of sense and recognition [No. 1108]

Computer Aided Manufacturers International (CAM-I) in the USA defines the humanoid aspects of the industrial robot as

a device that performs functions ordinarily ascribed to human beings, or operates with what appears to be almost human intelligence.

Another suggestion defines a robot in its developed form as

an automatic machine with a certain degree of autonomy, designed for active interaction into the environment.

By this definition a robot is an automatic machine, and perhaps this is the area where most confusion arises. A machine, to be accepted as a robot, must be able to respond to stimuli based on information received from the environment, no matter how restricted that environment may be. The robot will interpret the response either blindly, or by active sensing to bring about all the changes required in its environment. In doing so, decision making and performance of the activities defined in the programs are the robot's functions. All other functions can be called co-functions.

For further clarification of what is or will be the true nature of a robot, we separate the functions of the robot operation in three areas: 'sensing' the

---

*This denotes the reference number in the 'Glossary of terms for industrial robots', *JIS* B 0134-1979 (see Appendix 1.B).

environment by external perception sensors, e.g. vision, voice, touch, proximity, etc., 'making decisions' about the information received from the sensors, and 'deciding' the action required if a condition does or does not exist. Such robotic experience based on these decisions may be kept and used in a semantic manner for learning to react to similar events in the future. This will lead to robotic expert ability where the robot can operate proficiently without human interaction.

1.3   HUMAN ARM CHARACTERISTICS

In essence the robot manipulator resembles the human arm in appearance, structure, and in many of its functions. It is therefore worth considering briefly some of the human arm's most important characteristics.

The human arm consists of two distinct parts: the wrist with three minor joints, and the arm's two major joints, that is the shoulder and the elbow. The function of the human wrist is to provide the orientation of the object held by the hand. Its basic performance specification may be defined as follows: hold your right arm and hand straight out, keeping the palm in downward direction; this is the reference angular position (0°). Then rotate your wrist as far as you can in both a clockwise and anticlockwise direction. This is the roll motion and its possible limits are at $-180°$ and $+90°$ respectively, i.e.

$$\textbf{ROLL}(angle) = 180° \leftrightarrow 0° \leftrightarrow +90° = 270°$$

Again holding the right arm in the reference position (0°) and without rolling the hand, move the wrist from the initial straight position as far as possible in a downward and then in an upward direction. This is the pitch motion and its limit positions are at $-90°$ and $50°$, i.e.

$$\textbf{PITCH}(angle) = -90° \leftrightarrow 0° \leftrightarrow +50° = 140°$$

Holding your right arm straight out and the wrist making neither a roll nor a pitch motion, let the fingers point horizontally as far as possible to the right and then to the left. That is the yaw motion and its limit positions are at $-45°$ and $+15°$, i.e.

$$\textbf{Yaw}(angle) = -45° \leftrightarrow 0° \leftrightarrow +15° = 60°$$

Roll, pitch and yaw are considered independent motions and therefore referred to as degrees of freedom (see Section 1.8). The robotic interpretation of the wrist motions will be dealt with in detail in Chapter 3.

The second part of the human arm has two major joints with three degrees of freedom, two in the shoulder and one in the elbow. But the robot has a shoulder with one degree of freedom only, and thus the robot waist is introduced as a substitute for one of the shoulder's independent motions. In the case of a human the waist has a different function, i.e. it is for maneuverability by counterbalancing the human body against the gravitational load in changing body postures. This kind of counterbalance for robots is sometimes used by applying a mechanical loop in its structure.

At the end of the arm there is the hand with five fingers. Each finger has three degrees of freedom. If the fingers are used for gripping in any configuration, the finger joints are not independent and they are not involved either in positioning or in setting the orientation of the object held by the hand. If the fingers are used individually and independently, they offer degrees of mobility for motion dexterity (see Section 1.8). On the other hand, the robot gripper very rarely has independent mobility and consequently no dexterity. It is worth commenting on the applicability of the fingers, *viz.* that in 85% of pick-and-place types of industrial work only the thumb, first and middle fingers are used. This is the reason why in the case of a robot the gripper does not yet have more than two or three fingers.

Our further concern is the arm's articulation and, to a lesser extent, the leg's locomotion. Both arm and leg can be simulated as mechanical systems, as shown in their macroscopic mechanical models in Figure 1.1.

Regarding the arm's dynamic performance, we refer only to the distinct structural resonances occurring along the length of the arm and hand. Knowledge of the structural resonant frequencies of the human arm provides useful dimensional and operational information for robot arm kinematic design (see Section 7.3). It is interesting to note that the frequency bandwidth of the human arm from the shoulder to the hand is exponentially increasing and covers approximately three octaves [1.2].

One of the important features of the arm's structure is the ratio of the length of the upper arm to that of the forearm, which is around 1.2. This means that the robot's forearm should be equal to or slightly shorter than the upper arm. If this is not satisfied, as indeed it often isn't, then there will be a robot performance impairment, as shown theoretically in Chapter 6. When looked upon as a mechanical system, the human arm can be considered as a large scale hierarchical system composed of a number of linear and nonlinear elements (see Section 7.5).

1.4   MASTER–SLAVES AND ROBOT MANIPULATORS

In this section we are concerned only with a few representative examples of early manipulators, developed one or two decades ago. In the following,

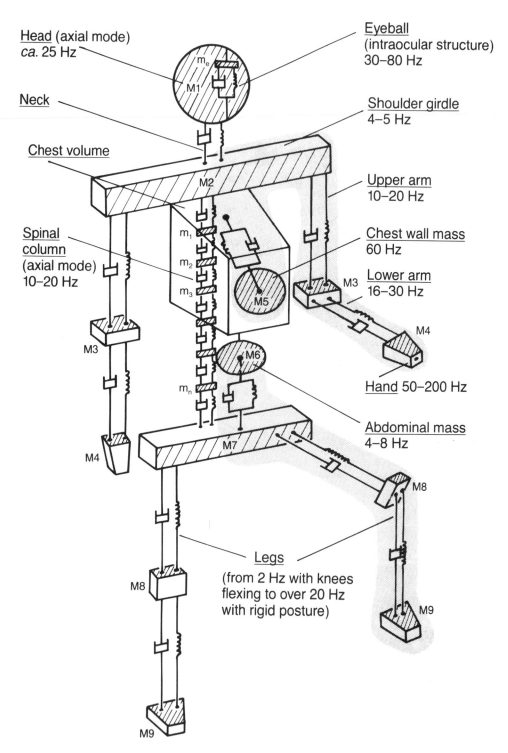

Figure 1.1 Simplified mechanical system representing the human body's kinematic geometry with particular emphasis on the upper limbs.

therefore, we will first describe master–slave manipulators and then a few early industrial robot manipulators of today.

Master–slave manipulators

The forerunners of industrial robots are human-operated master–slave manipulators. The 'master' is manipulated by a human operator and the 'slave' replicates the movements of the master. They are characterized by two manipulators either in direct mechanical linkage or connected together by electrical means. The latter could mean also a telecommunication link for very long distances in which case these systems are called teleoperators. Many different types of master–slave manipulator were and still are in use in both industrial research and development laboratories for handling radioactive and other dangerous materials [1.3]. For such work and for space research the master–slaves are indispensable.

This type of master–slave manipulator may have two modes of operation, either unilateral or bilateral. In the unilateral mode only the master is activated and the slave replicates the motion but does not report back to the master any unexpected event. In the bilateral mode both ends of the manipulator can be activated. Thus, when the slave manipulator encounters an obstacle in its own environment, the human operator not only sees the obstacle but also feels it and counteracts accordingly by activating the efferent nervous system* of the operator in an autonomous manner (reflexes). This operation is based on feeding back signals to the operator via a closed loop system. Examples of bilateral master–slave manipulators are shown in Figures 1.2 and 1.3.

It became feasible, that by removing the efferently activated human arm and the afferently responding human brain from the master–slave manipulator and by augmenting the slave manipulator under computer control, the device became a new type of manipulator, the robot.

Representatives of robot manipulators

Despite the fact that more than one hundred robot manufacturers are listed in various yearbooks [1.4], we deal here with only a few of the first manipulators to appear on the market. One of the first industrial robots was the Unimate-2000 series manufactured by Unimation Inc., shown in Figure 1.5. This robot appeared in the early 1960s and was introduced onto automobile assembly lines resulting in a great increase of productivity. Another early Unimation robot is the Programmable Universal Machine for Assembly, briefly called

---

*The efferent nervous system transmits the decisions from the human operator's brain to actuators [1.3].

†The afferent nervous system conducts signals from the external world via sensing devices [1.3].

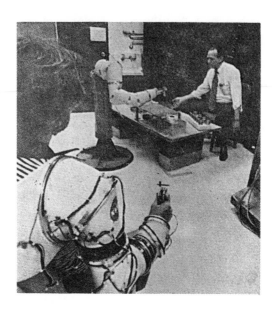

Figure 1.2   Human activated NASA-Ames master–slave manipulator, developed by Ames Research Center of the National Aeronautics and Space Administration, California.

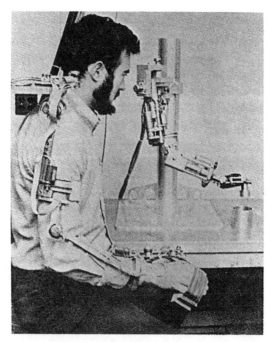

Figure 1.3   Human activated Rancho master–slave manipulator developed by Rancho Los Amigos Hospital, Los Angeles.

PUMA. The best-known members of this robot family are the five and six axis PUMA-550 and PUMA-560* robots respectively. The five axis PUMA is shown in Chapter 4 in Figure 4.1.

Another famous industrial robot was introduced by Cincinnati Milacron in the mid 1970s, shown in Figure 1.4. This robot became known as 'Tomorrow's Tool Today' or just T^3, and is a computer-controlled general-purpose machine. The new version of Cincinnati Milacron is the T^3R^3, which has the same major joints as Milacron's T^3 robot but incorporates a new multi-roll structured wrist. This new wrist on the T^3R^3 robot is built in a spherical geometry and has three roll axes coincident in the center point of a sphere. The two roll axes closest to the flange on the forearm, working in combination, provide a pitch–yaw motion, each having 230° rotation. The third axis provides continuous roll for the rotation of the tool. So the result is a pitch–yaw–roll configuration (see Figure 3.3).

The third early robot is one of the best-known, Asea's IRb-6 manipulator. It hasn't taken long for it to become one of the most popular robots in industry.

1.5 CLASSIFICATION OF ROBOTS

Robot classification may be considered on the following basis:

(1) Structural configuration and robot motion
(2) Trajectories based on motion control
(3) Performance characteristics of the robot.

In the following we will deal with the first two classifications. The third is probably the most important but it is beyond the scope of this book because it requires higher-level kinematics and kinetics considerations.

Classification based on structural configuration and robot motion

For the purpose of robot classification based on structural configuration and robot motion we distinguish between three basic motions in their operation. The first is a rotation about the longitudinal axis of a link between two joints which is called swivel motion. The second is also a rotation about the transverse axis in the joint which is called bending motion. The third is a linear motion in the direction of the longitudinal axis, either extensional or constructional, which is the prismatic motion. According to a robot's joint movements there are the following well distinguished basic robot configurations:

---

*PUMA is a registered trademark of Unimation which is now a Westinghouse Company.

   (i) Revolute (jointed arm) robot (Figure 1.4)
  (ii) Polar (spherical) robot (Figure 1.5)
 (iii) Cylindrical robot (Figure 1.6)
 (iv) Cartesian (rectangular) robot, sliding-type (Figure 1.7)
  (v) Cartesian (rectangular) robot, gantry-type (Figure 1.8)
 (vi) SCARA-type robot (Figure 1.9).

(i) *Revolute (jointed arm) robot*

Revolute robot is the type which best simulates a human arm, and is often referred to as an anthropomorphic robot. Because of this it is more easily adapted to an existing human workstation than any other type of robot. The revolute robot consists of three major rotary joints acting as the waist, the shoulder mounted on the waist, and the elbow mounted at the end of the shoulder link (Figure 1.4). Typical examples are Asea (IRb-6, Sweden), Unimation (PUMA 550, 560, 760, US and UK).

(ii) *Polar (spherical) robot*

This robot rotates about the axis of its waist on the base. The second axis is a horizontal rotary joint, allowing the arm to rotate in a vertical plane. Making use of both axes, the arm can sweep through a partial sphere. This mathematically corresponds to a polar coordinate system, thus this kind of robot is classified as polar. The third degree of freedom is provided by a prismatic joint built into the arm which allows it to move in and out. The robot can sweep through partial spheres of radii depending on the length of the prismatic joint, shown in Figure 1.5. Typical examples are Unimation (Series 1000, 2000, 4000, US), Prab (Model 4200 and 5800, Norway), GEC (Little Giant, UK).

(iii) *Cylindrical robot*

This robot consists of a base, a horizontal arm and a prismatic joint built into the horizontal arm shown in Figure 1.6. The whole base can move up and down. The horizontal arm swivels around the vertical column, describing a partial cylinder in space. This mathematically corresponds to a cylindrical coordinate system, thus this kind of robot is classified as cylindrical. The prismatic joint built into the horizontal arm can slide in and out, remaining parallel to the base. Typical examples are Prab (Model E, FA, FB, FC and Model G Series, Norway), Paterson (Wiper 2500, 3000, UK) Fanuc (M-M1, M-M2, M-M3, Japan).

(iv) *Cartesian (rectangular) robot, sliding type*

There are three perpendicular traversing axes, realized by an up/down, a left/right, and a forward/backward moving prismatic joint. This mathematically corresponds to a cartesian coordinate system, thus this kind of robot is

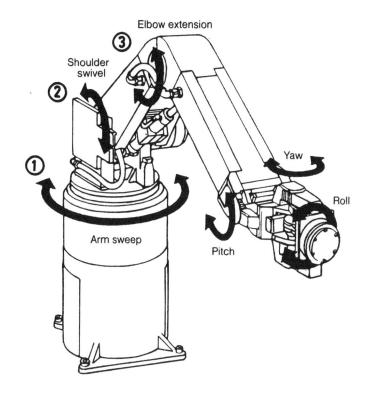

Figure 1.4 Revolute robot (Cincinnati Milacron T³). The body rotates about the longitudinal axis in a vertical direction. The shoulder and elbow joints have transverse axes and their movements are in the vertical plane. Courtesy of Cincinatti Milacron.

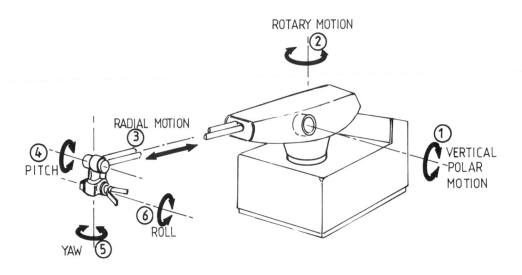

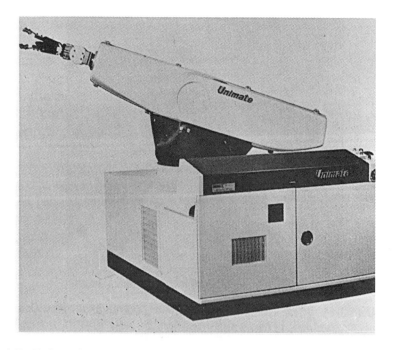

Figure 1.5   Polar robot (Unimate 2000). The body rotates about the longitudinal axis in a vertical direction. The shoulder rotates abouts its transverse axis and produces vertical polar motion. The arm link extends in its axial direction, i.e. radial motion. Courtesy Unimation, UK.

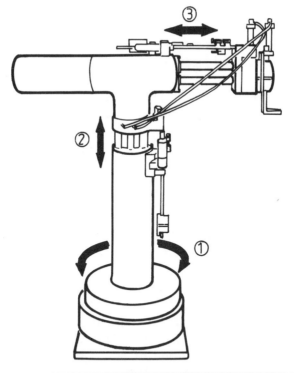

Figure 1.6 Cylindrical robot (Paterson Wiper 2500). This robot has two prismatic motions and one rotation about its vertical axis. Courtesy Paterson Production Machinery Ltd.

classified as cartesian (Figure 1.7). Despite the fact that this robot is of high precision, it is not preferred for many applications because of its difficult adaptability to the existing human-operated workstations. A typical example is DEA robot (Digital Electronic Automation SpA, Model Pragma A3000, Italy).

(v) *Cartesian (rectangular) robot, gantry-type*
This type of robot, shown in Figure 1.8, has the same structure as the sliding-type cartesian robot. The only difference is that it has a gantry for keeping the robot in sliding operation. A typical example is IBM (7565, US).

(vi) *SCARA-type robot*
New robot kinematic configurations can be obtained by combining the properties of the basic robot representatives outlined above. For instance if the revolute and cylindrical robot kinematics are combined, the result will be a new type of robot called SCARA, where SCARA stands for Selective Compliance Assembly Robot Arm (see Figure 1.9). Its rotary joints have vertical axes, allowing movement in a horizontal plane, which corresponds to both revolute and cylindrical coordinates. The SCARA configuration has vertical major axis rotations for which the gravitational load, Coriolis and centrifugal forces do not stress the structures as much as they would if the axes were horizontal. This advantage is very important at high speeds and high precision. The best examples are IBM (7535 and 7545, US), Meta Machines (adept one, UK).

Classification based on path control

There are two basic forms of robot path control:

  (i) Point-to-point (PTP) control
  (ii) Continuous path (CP) control

With point-to-point control the robot is programmed to pause at each point to plan the next step in a predetermined manner. Despite the fact that the motion is not controlled between the set points, it usually occurs along a natural path, depending upon the kinematic geometry of the robot. On the other hand the robot under continuous control can follow any arbitrary path accurately. A point-to-point controlled robot offers greater precision in terms of accuracy and repeatability. The continuous path control results in a smoother movement along the defined trajectory but there is a speed penalty, which is a function of the step sizes computed by the master computer in real time using interpolation methods. The penalty may be a 15–25% speed reduction, resulting in uneconomic control of the process, i.e. the efficiency will be lower compared with the same robot operating in point-to-point control mode.

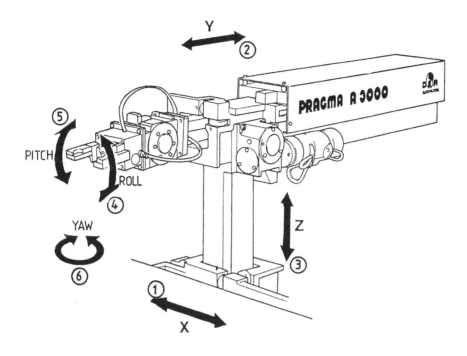

Figure 1.7 Cartesian coordinate robot (Pragma A3000 DER). This robot has two perpendicular axis movements in both the horizontal (x, y) and vertical (x, z) planes. Courtesy of Digital Electronic Automation SpA, Italy.

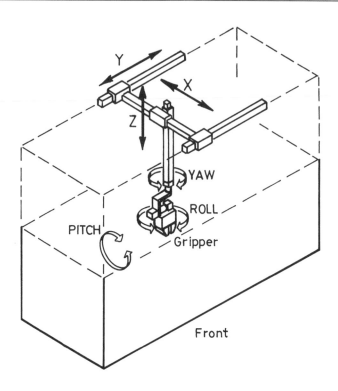

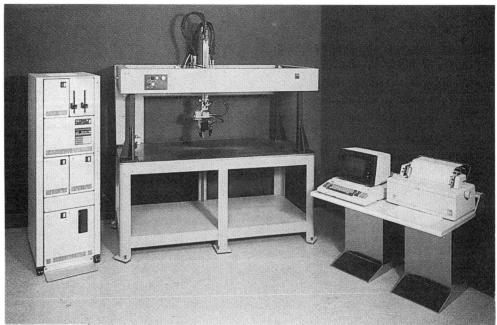

Figure 1.8 Cartesian coordinate robot, gantry-type (IBM 7565). Three translatory motions in x, y and z directions. Courtesy IBM.

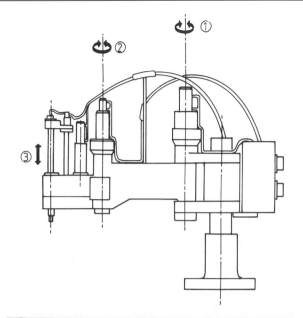

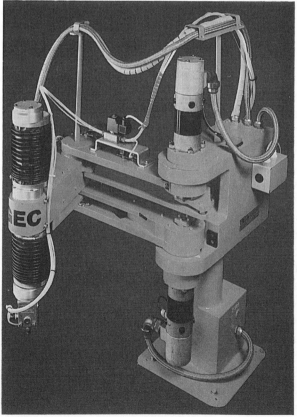

Figure 1.9 SCARA-type robot (GEC A3020, Selective Compliance Assembly Robot Arm). The shoulder and elbow joints can rotate in the horizontal plane about their vertical axes. At the end of the robot there is a vertical axis for lifting. Courtesy GEC Electrical Projects Ltd.

1.6   ROBOT SYSTEM ORGANIZATION AND OPERATION

To understand what kind of kinematic structure a particular robot has, how the robot works, and how a robot best fits into a workspace, involves knowledge from mechanical and control engineering, analog and digital electronics, computer science, manufacturing process and management science. Anybody who studies robotics will need some supporting knowledge in all these disciplines.

Today's industrial robot systems, while varying in size, shape and performance, are generally composed of five main groups of component: manipulator, controller, power drive including the power supply, teach pendant or manual control either with or without monitor, and the auxiliary peripheral terminals [1.5] as shown in Figure 1.10.

The robot manipulator consists of mechanical devices that do the work and provide the mastery of articulation, powered by pneumatic, hydraulic or electrical drives. As many drives are required as there are kinetmatically independent motions. Their number is usually six but not all applications require that many. For instance, a workpiece transfer may only require three axes, pick-and-place four axes, whereas paint spraying requires a minimum of five axes, and assembly work six.

Power amplifiers drive the axis motors, while the controllers accept the decoded software information. The various modes of motion and their sequential points along the path are stored as programs in the memory, located usually in the robot's custom-built computer. It is the controller that coordinates the motions about the manipulator axes and interacts via I/O devices with both interoperceptive and exteroperceptive sensors* and synchronizes external machines, devices, conveyors or tools with which the robot carries out the work, test and inspection. Controllers vary in complexity, and may be as simple as logic circuitry or as sophisticated as 8- to 32-bit mini- or microcomputer systems [1.6].

There are various types of computer devices for receiving, processing and transmitting programs and data, developing the user's software, editing programs, servicing the robot, etc. There are also various types of storage facilities for saving the necessary information. The memory in the controller is of two types: nonvolatile and volatile. Nonvolatile memory keeps the information and is not erased when the power is off. Volatile memory is one which is erased when the robot system is powered down or the power switched off. All robot systems have both volatile and nonvolatile memories.

The nonvolatile memories, the Read Only Memory (ROM), the Programmable Read Only Memory (PROM), and the bubble memory keep all the

---

*Abbreviations stand for internal and external perception sensors.

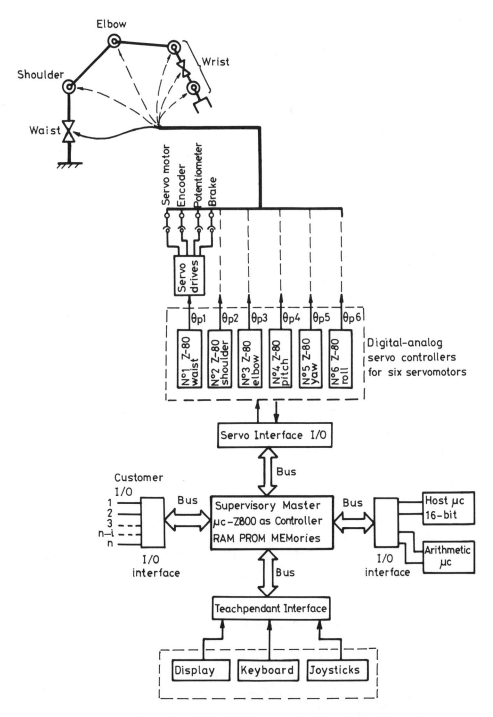

Figure. 1.10 Operational block diagram of a six-axis revolute robot.

information about how the robot system works. The volatile memory, called Random Access Memory (RAM) or Read/Write Memory, is where the application programs, the arithmetic and logic calculations, or even the procedures of high-level mathematical operations are preserved. It is from RAM that the controller gets the continuous data for controlling the robot via the axis motion. In order to keep more data to hand, there is usually back-up storage such as floppy disk, winchester disk or magnetic tape.

1.7   PRINCIPLES OF STANDARDIZATION AND INDUSTRIAL ROBOT STANDARDS

In this section we will deal with only a few aspects of standardization and discuss some standards that already exist. We note that no agreement, either nationally or internationally, has yet been reached on the definitions and component designations of the robot and robot interfacing devices. Until that happens the definitions in the specifications and notations in the figures will be based on the best suggestions found in the literature and various international committees' drafts and proposals.

General aspects in robot standardization

There are two distinct categories of standardization for industrial robots: the terminology and the characterization of robot components. Others may follow later, such as the rules of robot design, testing and measuring methods, inspection and specification, etc. The first two need to be established as early as possible; the others should be delayed until the corresponding disciplines are better established.

The general view is that standardization is a very important factor of growth, but its early introduction withholds progress particularly during the development stage. Note that robots are still very much in a volatile period. Despite this fact manufacturers and robot users are eager to know the state of standardization and the proposed standards in robotics and robotization. There are certainly many conflicting definitions as well as a variety of standards. In the current interim period a six-level grouping of possible robot manipulators standardization areas is proposed for consideration [1.7]:

(1) Terminology and symbology
(2) Characterization with specifying methods
(3) Standards of robot construction
(4) Design rules
(5) Testing methods and test codes
(6) Performance inspection.

Terminology and symbology in robotics represent the fundamental concepts which define the meanings of robot items. They are essential for describing methods of designing, manufacturing, testing, measuring, data processing, and programming of the robots and their ancillary systems. All these may briefly be considered under characterization. Standards of construction detail the characteristics of the items and components that are required in designing robot mechanisms and their applications. Robot kinematics design and testing rules are similar to the standards of mechanisms that already exist in practice although there are a number of differences as the various existing guidelines already indicate. Performance inspection is not yet sufficiently advanced to set its standards.

The great difficulty in robot standardization is that, by its very nature, it cuts across many technological and engineering fields. Whether one is concerned with the standards of robot technology or with robot metrology, a wide range of expert knowledge is required. Engineers, mathematicians, computer scientists and representatives of many other disciplines involving both practical and theoretical aspects can make important contributions throughout the whole spectrum of robot standardization.

Standardization is feasible primarily through international exchange of research and development by disseminating technological information on robotics. Initiative and organization of such events would primarily be centered around international conferences and congresses, but the leading role in standardization would be left for the time being under the aegis of the national robot institutions. No doubt the importance of standardization in robotics and in all associated disciplines will considerably increase in this decade and the national robot institutions will play a great part in it. For this reason a list of all existing national robot associations and societies is provided in Table 1.A.1. in Appendix 1.A [1.8].

Computer users working in CIM (Computer Integrated manufacturing) industry encounter great difficulty when they have to interconnect local automation systems. Whatever communication systems, standards or protocols are chosen for interconnections, even more difficulty is encountered when it comes to diagnostics. The neglect of standardization has already caused enormous problems for the manufacturer and even more for the end-user. It is enough that the end-user should have to go to several suppliers to get equipment that can be interconnected readily by a broad-band communication network, but when there are problems interconnecting the equipment on the factory floor because of hardware and software incompatibility, then the situation becomes unacceptable.

Lacking standards, many manufacturers have had to introduce their own terminology and characterization for computer and robot functions. Suppliers and users of FMS (Flexible Manufacturing Systems) must be aware of the current initiatives being taken on standards for computer communication, including robot interfacing schemes, by General Motors* in US. The result is the MAP (Manufacturing Automation Protocol). MAP is a set of protocols for intermachine connection under which computer controlled factory equipment from various manufacturers can communicate with each other on a common communications network. Although the current version MAP 2.1 is under further development, a more advanced version of MAP (MAP 3) is already in the offing.

A companion protocol to MAP is the TOP (Technical Office Protocol) sponsored by Boeing Corporation in US. TOP is a complementary communications specification for use in technical and office environment. Functions that TOP will address include: document, spreadsheet and graphics exchange; print, plot, file and directory servers; electronic mail, and store and forward messaging; file transfer and distributed database interfaces.

Both MAP and TOP have already gained the support of leading manufacturers in the US and Europe as well [1.9], [1.10]. In fact MAP and TOP has been officially combined into one MAP/TOP Users' Group† and became an official Technical Group of the Society of Manufacturing Engineers (SME) in North America in 1985. Similarly a European MAP organization was formed under the name of European MAP Users Group (EMUG)‡ in 1986. MAP/TOP linked systems enable effective communication between 'islands of technology', such as machine tools, computer aided design installation, robots or engineering workstations on the shopfloor.

Industrial robot standards

The drafting and proposing of industrial standards by the national robot associations and societies is everywhere in progress. In such respect the Japanese Industrial Standard (*JIS* B 0134, 'Glossary of terms for industrial robots') [1.1] and National Bureau of Standards (NBS), US, Department of Commerce (Special Report 459) are worth mentioning because those are the most advanced proposals for terminology standardization. The former draft proposes the following groupings of terms in robot terminology:

(1) Types of industrial robot

    (1.1) classification based on information and teaching
    (1.2) classification based on sequence information
    (1.3) classification based on operating configuration

---

*General Motors Technical Center, Manufacturing Building, A/MD-39, 30300, Mound Road, Warren, Michigan 48090-9040, USA.
†The MAP/TOP Users' Group, the Society of Manufacturing Engineers, one SME Drive, PO Box 930, Dearborn, Michigan, 48121, USA.
‡EMUG, Cranfield Institute of Technology, College of Manufacturing, Bldg 30, Cranfield, Bedford MK43 0AL, UK

(2) Basic terms relating to functions
(3) Basic terms relating to performance and characteristics
(4) Terms relating to work performed
(5) Terms relating to operation and control
(6) Terms relating to intelligent robots
(7) Terms relating to safety.

To ease the problems due to lack of standards the authors suggest the use of currently available glossaries of terms. We therefore provide the most important terms from the *JIS* B 0134-1979 in Appendix 1.B, selecting some items from each grouping above (see Tables 1.B.1 to 1.B.7 in Appendix 1.B).

One difference in standardization between Western and Japanese approaches is that the West does not classify the fixed (dedicated) machines as robots, although sometimes the dedicated machinery is quite flexible. For instance, a cutting machine automatically executes a cutting sequence which cannot be changed regardless of how the input is organized. By Japanese standards this cutting machine is a robot (e.g. Japanese term: fixed sequence robot, *JIS* B 0134-1979, No. 1104). Another example is a pick-and-place device which picks up, say, a sheet from a position and then drops it into a cutting machine. Despite the fact that the machine can physically be adjusted, i.e. it is flexible, by Western standards it is not a robot. It would be considered a robot only if the adjustment of the operation could be done by software changes. But this kind of pick-and-place machine, by Japanese standards, is a robot.

To draw a distinction between robots and computer numerical control (CNC) devices is even more difficult. Despite the fact that such machine tools are controlled by computer, using prerecorded programs, and the software is adjustable, by Western standards they are not considered flexible enough to be termed robots, because their CNC properties are dedicated to a fixed job. But on the other hand they are robots by Japanese standards.

**1.8   INTERPRETATION OF DEFINITIONS AND
         CONCEPTS IN ROBOTICS**

Kinematics, kinetics and dynamics

Kinematics deals with motion in isolation from both the forces and torques associated with linear and angular motions respectively. Furthermore the successive derivatives of motion with respect to time, namely velocity and acceleration, and even higher order derivatives of motion are all combined into kinematics [1.11]. In short, kinematics in robotics may be interpreted as the subject of mechanical movements of all kinds. On the other hand kinetics involves forces, torques, energy, moment of inertia, mass, equilibrium, stability etc. Kinematics and kinetics are together combined into dynamics.

Degree of freedom (DoF) and degree of mobility (DoM)

The number of independent movements that an object can perform in a three-dimensional space is called the number of degrees of freedom (DoF). A rigid body when moving freely in space has six degrees of freedom, three for location and three for orientation. These two kinds of independent movement, shown in Figure 1.11, are

- (i) the three translations T1, T2 and T3, which represent the linear motions along the x, y and z axes respectively,
- (ii) the three rotations R1, R2 and R3, which represent the angular motions about the x, y and z axes respectively.

By using three orthogonal translations and three rotations about the orthogonal axes, the state of an object, i.e. its location and orientation anywhere in the workspace of the robot, can be completely defined. This might seem to suggest that a robot should have no more and no less than six independent degrees of freedom.

In fact, however, there are robots in use having fewer or more than six joints. Robots with fewer than six obviously have constrained motion, but

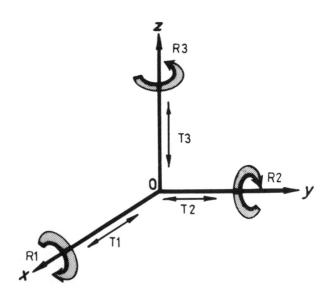

Figure 1.11 Representation of six degrees of freedom. (*a*) The three translatory motions (T1, T2, T3). The three rotary motions (R1, R2, R3).

there are many jobs which can be done by robots having five or even four joints only. A robot with more than six has surplus joints which may enhance its performance by improving its motional dexterity. Nevertheless, in most cases the surplus joints tend to cause programming difficulties in calculating coordinate frame transformations. However, which joints are surplus depend on the robot's kinematic configuration in operation. Whether we use a six, or a more or less than six, jointed robot depends upon the application in hand.

Note that any rigid object moving freely in three-dimensional space has six degrees of freedom, i.e. $2n$ where n is the number of dimensions. But in a two-dimensional space an object has three degrees of freedom: two translatory and one rotary. For instance, a straight line object of fixed length in two-dimensional space has only $2n-1$ degrees of freedom (and not $2n$).

It is essential to distinguish between kinematically independent and dependent arm joints. As far as the latter, i.e. the redundant joints, are concerned we will say that each one contributes a degree of mobility (DoM). Every joint, either revolute or prismatic, represents a degree of mobility, but not necessarily a degree of freedom. The schematic interpretation of DoF and DoM in kinematic terms is shown in Figure 1.12 where it can be seen that there are some cases when the same number of joints, i.e. the same number of degrees of mobility, results in a different number of degrees of freedom.

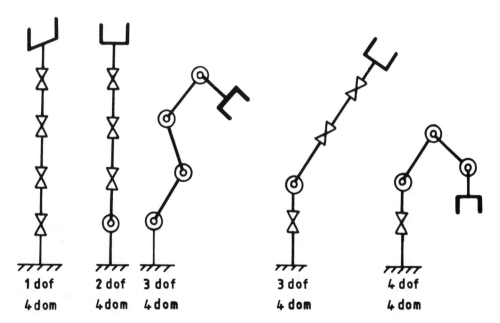

Figure 1.12 Schematic interpretation of degree of freedom (DoF) and degree of mobility (DoM). Longitudinal axis rotation ⦻ . Transverse axis rotation ⊚ .

Accuracy and repeatability

Accuracy relates to a robot's ability to move to a command position at a specified velocity within its established working area. The position and/or velocity is usually measured at the end of the arm. When the robot is used to manipulate a tool, accuracy is measured at a point that represents the furthest extension of the tool. Accuracy is the difference between the measured value and the command value of a specified position in the robot's workspace. Repeatability is a measure of the spread of positions in a series of attempts to position the manipulator at a fixed location [1.12]. A graphical illustration of accuracy and repeatability is shown in Figure 1.13.

Position accuracy is measurable by making a series of measurements of the difference between the actual position and desired (programmed) position of the manipulator. Good repeatability is the ability to repeat the same position several times within a specified tolerance. As this repeatability specification is given for the majority of robots in this way, we may suggest that repeatability is a more important consideration than accuracy. Nevertheless, if a robot has good repeatability but is not accurate or vice versa than it is of little use in some applications such as welding, painting, assembling, etc.

In order to provide further clarification of the various positioning accuracies, we will first deal with the relationship between robot operation and the corresponding errors in terms of accuracy. For this purpose we introduce a model representing the required, taught, and actual positions. Between them there are the related accuracies, teaching accuracy, playback accuracy and positioning accuracy, as shown in Figure 1.14.

Ambiguities with robot velocities

When defining control performance in robot operation, speed is a very vague concept and even more deceptive than position accuracy. Hence let us first consult the various speed characteristics which in their simplified presentation may help the interpretation of various path velocities shown in Figure 1.15(*a*). Despite the fact that curve 2 has higher maximum speed than curve 1, its displacement is shorter as shown in Figure 1.15(*c*). In fact curve 2 has no slewing velocity $\theta_{v,s}$ because the robot at maximum speed begins decelerating to its terminal position. Consequently the maximum speed is not a sufficient parameter to specify the robot's motion. A knowledge of the acceleration profiles is also essential (see Figure 1.15 (*b*)).

It is also shown that the velocity control may include a step down deceleration near to the target point as shown on all the curves of Figures 1.15 (see points B→E), and this may be adjustable. The characteristics of velocity and acceleration are of paramount importance to estimate the working time

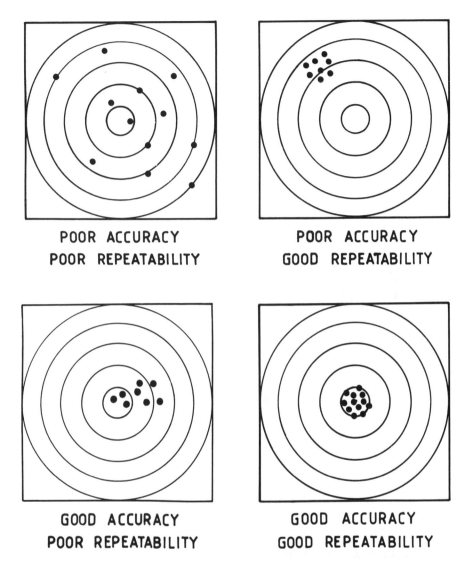

POOR ACCURACY
POOR REPEATABILITY

POOR ACCURACY
GOOD REPEATABILITY

GOOD ACCURACY
POOR REPEATABILITY

GOOD ACCURACY
GOOD REPEATABILITY

Figure 1.13   Accuracy and repeatability. Courtesy of *Robotics Age*.

and cycle time for manufacturing various products and in the calculation of the efficiency and productivity [1.7].

Long-term speed holding accuracy

The control for robot drive has traditionally used analog techniques. Analog systems are easily understood and measured, but suffer from the fact that both position and speed accuracy depend on features which are independent of the

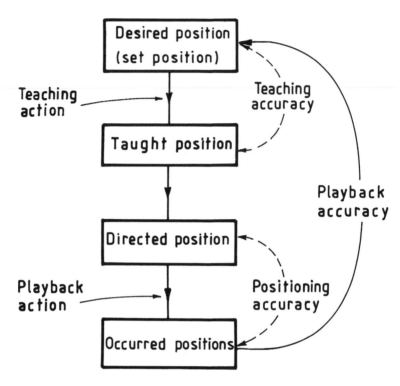

Figure 1.14 Model of positioning sequence operation and the related accuracies.

control loops. The best analog systems can produce almost 0.1% accuracy. The adoption of digital techniques can provide control systems where long-term position accuracy and speed holding is almost absolute. A digital system using sampling techniques to measure speed would typically produce long-term speed holding accuracy better than 0.01%.

Comment on robot intelligence

Existing robots cannot yet organize what they sense into an action, even in the most coherent environment conditions. The goals of the current robot are entered externally by setting the information in coded form, but how to interpret the perceived information for action and how to implement that action must be determined in advance in the software by the designer.

The great majority of the present robots are not capable of investigating their own environment. But most of the current research in robotics is tending in this direction, promoting progress towards the next generation robot era. The intelligent robots will be able to modify and even extend their stored knowledge by using an appropriate learning and decision-making algorithm. The future generation of robot will be able to act on knowledge and concepts

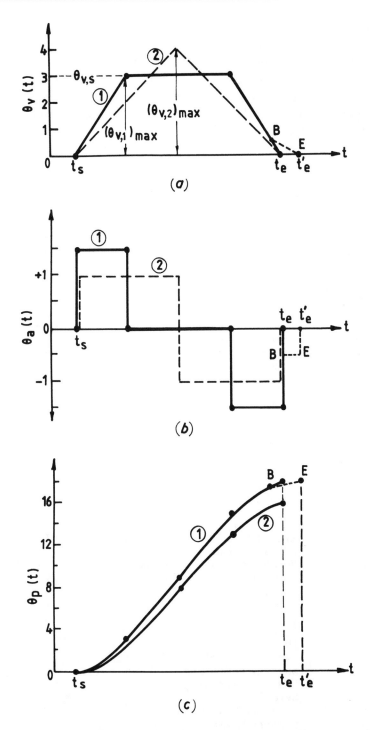

Figure 1.15 Motion models of trajectories. (*a*) Velocity profiles. (b) Acceleration profiles. (*c*) Displacement (position).

processing, based on a new 'predicate' logic theory. This possesses potentially the capability to formulate a clear picture about the robot's environment by receiving more sophisticated afferent information.

A robot, to function intelligently, will receive a continuous flow of information at very high speed, which includes some previously gained expert experience stored in the memory and in addition a continuous supply of ever-changing inputs from its environment. Both knowledge and information obtained by internal and external sensory perception will provide the artificial intelligence within the program to control the robots of the future.

1.9   APPENDICES

Appendix 1.A Robot associations and societies throughout the world

Table 1.A.1

Number	Name and address	Established
1	Japan Industrial Robot Association (JIRA) c/o Kikaishinko Kaikan Bldg. 3-5-9 Shibakoen, Minato-ku, Tokyo 105 Japan	October 1972
2	Robotic Industries Association (RIA)* One SME-Drive, 20501 Ford Road PO Box 1366 Dearborn, Michigan 48121 USA	June 1974
3	Italian Society for Industrial Robots (SIRI) c/o Etas Kompass Via Mantegna, 6 Milano Italy	1975
4	British Robot Association (BRA) 35–39 High Street, Kempston, Bedford MK42 7BT United Kingdom	1977
5	French Industrial Robot Association (AFRI) 91 rue Falguiere, 75015 Paris, FRANCE France	1978
6	Robotics International of SME One SME Drive, PO Box 930, Dearborn, Michigan 48128 USA	1980
7	Swedish Industrial Robot Association (SWIRA) Storgatan 19, Box 5506 S-114 85 Stockholm Sweden	1980
8	Australian Robot Association 9 Queens Avenue McMahons Point, Sydney, NSW 2060 Australia	1981

Number	Name and address	Established
9	Fachgemeinschaft Montagtechnik Habdhabungstechnik und Industrie Robot Postfach 710109 c/o VDMA D-6000 Frankfurt (Main) 71 West Germany	October 1981
10	Danish Industrial Robot Association (DIRA) c/o Technological Institute Div. of Industrial Automation Gregersensvej Postbox 141, OK-2630 Taastrup Denmark	February, 1982
11	Singapore Robotic Association 5 Portsdown Road, Off Ayer Rajah Road Singapore 0513 Republic of Singapore	June 1982
12	Robotic Society of Japan (address, see JIRA)	January 1983
13	The Robotics Society in Finland PO Box 5500331 Helsinki 33 Finland	March 1983

*Note that the Robot Institute of America was renamed in 1984.

Appendix 1.B   Glossary of terms for industrial robots; *JIS* B 0134-1979 (translated from original Japanese)

Since March 1974 the standardization of robot terms and functions is an ongoing study initiated and organized by the JIRA under the assignment of the Industrial Science and Technology Agency of the Ministry of International Trade and Industry (MITI). Several surveys were conducted through questionnaires sent to JIRA members and many researchers. The aim of the study is:

   (i) to classify industrial robots on the basis of their structures and functions;
  (ii) to define the technical terms relating to industrial robots;
 (iii) to define the graphical and letter symbols used in describing industrial robots.

The result is the Japanese document mentioned above from which some proposed terminology is quoted in Tables 1.B.1–1.B.7.

Table 1.B.1 Terms relating to types of industrial robots.

Number	Term	Meaning
1101	Manipulator	A machine which has functions similar to those of the human upper limbs and moves an object spatially from one location to another
1103	Sequence robot	A manipulator which progresses successively through the various stages of an operation according to the predetermined sequence, condition and positions
1106	Playback robot	A manipulator which is able to perform an operation by reading out stored information for an operating sequence, including positions and the like which it learned beforehand by being taken manually through the routine
1108	Intelligent robot	A robot which can determine its own behavior/conduct through its functions of sense and recognition
1202	Repeating robot	A manipulator performing an operation repeatedly, according to a prememorized work program

Table 1.B.2 Basic terms relating to functions.

Number	Term	Meaning
2017	Path control	A control which ensures that a given path is taken by means of information relating to the path
2018	Continuous path control	A path control where the entire route is specified. *Note*: Will include pseudo-continuous path control by interpolation
2019	Point to point control	A path control wherein along the route a finite number of points passed are specified

Table 1.B.3. Basic terms relating to robot performance and characteristics.

Number	Term	Meaning
3018	Maximum speed	The maximum value of speed manifested at a specified location, under specified conditions of use. *Note:* The particular conditions of use such as no-load maximum speed, or rated-load maximum speed, etc., are to be specified
3019	Acting time	The time measured between being given the directive to move to a target position up to the termination of the movement
3020	Positioning accuray	This is based on No. 601 in *JIS* B 0181 (Glossary of terms for numerically controlled machine tools). *Note:* Quantitatively this is expressed as an error which will include errors in the control system which drives the robot, and errors in the controlled mechanical axes
3021	Position repetition accuracy (repeatability)	This is based on No. 602 in *JIS* B 0191 Reference 1: the level of agreement between positions when performed under the same conditions, and by the same method; Reference 2: at times, the term 'reproducibility' may also be used
3022	Position playback accuracy	The agreement between the set values and that reproduced

Table 1.B.4 Terms relating to the work performed.

Number	Term	Meaning
4001	Workpiece	This is the name of the object dealt with by the robot
4002	Loading	This is the fitting or placing-on of the workpiece onto the machine which then performs the work of matching, assembling, transferring, and the like
4003	Unloading	This is the removal of the manufactured product or partially manufactured product, from the machine which performs the work of machining, assembly, transfer, etc., when the work is complete
4004	Palletizing	Arranging materials on a pallet according to fixed rules
4005	Depalletizing	Removing in sequence materials which have been arranged on a pallet

Table 1.B.5   Terms relating to operation and control.

Number	Terms	Meaning
5014	Sequential control	Control which progresses through successive control stages following predetermined routines or predetermined conditions
5015	Feedback control	This is based on No. 4 in *JIS* Z 8116 (Glossary of terms used in automatic control)
5016	Servo system	This is based on No. 17 in *JIS* Z 8116

Table 1.B.6   Terms relating to intelligent robots.

Number	Terms	Meaning
6001	Artificial intelligence	The ability to learn, the capability of abstract thought, to be able to draw inferences, reasoning, environmental adaptability and recognition, etc.
6002	Intellectual function	The ability to engage objects in the neighbourhood in the following actions; learning, judgment, recognition, adaptation, etc., through the medium of detection, memorizing, etc.
6012	Automatic programming	The conversion by the robot itself of instructions, for a particular operation or job, which are available in terms readily comprehensible to a human, into a form such that the robot can follow; reference: See No. 0135 in *JIS* C 6330 (Glossary of terms used in information processing)

Table 1.B.7   Terms relating to safety.

Number	Terms	Meaning
7001	Danger prevention	A function which ensures the robot does not produce a dangerous condition
7002	Danger zone	The zone in which one should not enter because a dangerous situation will be manifested at certain times
7003	Safety pole	An erect pole for limiting the runaway of the robot in the case of abnormal operation; generally speaking, it will be strong enough to withstand impact by the robot
7004	Safety barrier	A barrier, or the like, provided around the danger zone to prevent human entry into the said danger zone
7005	Backup system	Auxiliary functions or auxiliary machine(s) provided beforehand such that if a fault occurs in some functional unit or machine in constant use in the system, the operation of the system may still continue uninterrupted

1.10   REFERENCES

[1.1]   Japanese Standards Committee, Glossary of terms for industrial robots, *JIS* B 0134-1979, January 1979.

[1.2]   Rasmussen, G. Human body vibration exposure and its measurement, *Technical Review*, Bruel & Kjaer, No. 1, 1982.

[1.3]   Microbot, Inc. Operation of the 5-axis Table-top Manipulator, Model MIM-5, 453-H Ravendale Drive, Mountain View, CA 9404, 1982.

[1.4]   Cugy, A and Page, K. *Industrial Robot Specification*, Kogan Page, 1984.

[1.5]   Clarke, H. and N-Nagy, F.L. Powering the electric arm, *Electric Drives*, Fall 1983.

[1.6]   N-Nagy, F.L. Promoting awareness of industrial applications of microprocessors, *Electronics and Power*, Vol. 26, No. 6, 1980.

[1.7]   Inagaki, S. What is the standardization for industrial robots? *The Industrial Robot*, March 1980, pp. 46–9.

[1.8]   Yonemoto, K. International cooperation and trends in the field of industrial robot, *Robotica*, Vol. 3, 1985. pp. 165–70.

[1.9]   General Motors' MAP Technical Review Committee, *MAP Specification to MAP Network Users and Manufacturers*, March 31, 1985.

[1.10]   Department of Trade and Industry, The DTI MAP event, Executive Summary, *Computer Communications into the 1990s*, July 1986.

[1.11]   Hunt, K.H. *Kinematic Geometry of Mechanisms*, Oxford University Press, 1978.

[1.12]   Hadzima, J. Introduction to data acquisition, *Robotics Age*, Vol. 6, No. 10, 1984, pp. 12–18.

[1.13]   Foster, D. *Automation in Practice*, McGraw-Hill, 1968.

2

MATHEMATICS OF ROBOT MANIPULATION

2.1   HOMOGENEOUS COORDINATE TRANSFORMATIONS

The position and orientation of a rigid object can be described by a coordinate system firmly attached to that object. Providing a geometrical representation of an object is given, then it will be enough to define the position and orientation of the coordinate system for reconstructing the object at arbitrary places. For this purpose it is necessary to define the relative position and orientation of the coordinate systems in a concise and explicit manner, and the use of homogeneous transformations is one of the best and most advantageous methods [2.1], [2.2].

To familiarize the reader with homogeneous transformation, let us take a point anywhere in space as the origin O_0 of a base coordinate system. By using a vector k we can describe the position of another coordinate system of origin O_1 in the base system of origin O_0. The vector k provides only the position of the origin O_1 without indicating anything about the relative orientation of the second coordinate system in the first one. Their relative orientation is defined by a matrix of rotation R having three rows and three columns.

The elements of the rotation matrix R are obtained by the projection of the unitary axis vectors (e_1, e_2, e_3) of the coordinate system of origin O_1 to the coordinate axes of the base coordinate frame. The spatial relationship between the base frame and the shifted/rotated coordinate system is shown in Figure 2.1.

The unitary vector triad (e_1, e_2, e_3) can be used to determine the orientation of the translated coordinate system, for which the following vector and scalar relationships are always true:

$$e_1 = e_2 \times e_3 \quad \text{and} \quad e_2 \circ e_3 = 0$$

where $\times$ and $\circ$ stand for vector and scalar multiplication respectively and $*$ will denote both vector–matrix and matrix–matrix multiplications.

In conclusion, therefore, the vector product (cross product) between any two unitary vectors provides the third one and the scalar product (inner product) indicates the degree of parallelism between them. In fact the inner product gives the projection of one vector on the other, therefore three nonzero vectors are said to be orthogonal if

$$e_1 \circ e_2 = e_3 \circ e_1 = e_2 \circ e_3 = 0$$

So any coordinate set with unitary vectors on its corresponding axes having the property above is an orthogonal coordinate system.

Thus the rotation matrix R is an orthogonal 3×3 matrix, which consists of

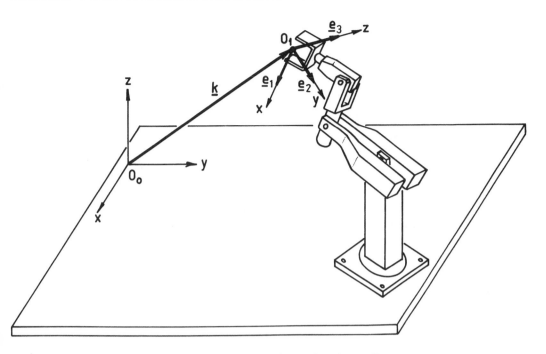

Figure 2.1 Spatial state of the base and translated coordinate systems.

the x, y, z components of the unitary vectors (e_1, e_2, e_3) projected to the coordinates of the base coordinate system.

$$R = \begin{bmatrix} e_{1x} & e_{2x} & e_{3x} \\ e_{1y} & e_{2y} & e_{3y} \\ e_{1z} & e_{2z} & e_{3z} \end{bmatrix}$$ (2.1)

Let us now find the coordinates of the point P in the base frame when its position is given with respect to the coordinate system of origin O_1 (see Figure 2.2).

The position of point P in the base coordinate system is calculated by premultiplying the vector p' by the rotation matrix R and adding to it the position vector k.

$$p = k + R * p'$$ (2.2)

The same result is obtained by using a single matrix multiplication, when the position vector $k(k_x, k_y, k_z)$ is also included in a combined matrix as the fourth column and 1 becomes the fourth element of the vector p' as shown in Equation 2.3.

$$
\begin{bmatrix} p_x \\ p_y \\ p_z \end{bmatrix} = \begin{bmatrix} e_{1x} & e_{2x} & e_{3x} & k_x \\ e_{1y} & e_{2y} & e_{3y} & k_y \\ e_{1z} & e_{2z} & e_{3z} & k_z \end{bmatrix} * \begin{bmatrix} p'_x \\ p'_y \\ p'_z \\ 1 \end{bmatrix}
$$

(2.3)

By adding a fourth row, consisting of three zeros and a one to the 3×4 matrix in Equation 2.3, a 4×4 matrix can be constructed which is described as a homogeneous transformation matrix and is given in Equation 2.4. A homogeneous transformation matrix combines the position vector k with a rotation matrix R to provide a complete description of the position and orientation of a second coordinate system with respect to the base frame. When transforming the coordinates of a point P from one coordinate system to another, the base (absolute) and the other (relative) coordinates are shown, as in Figure 2.2, and the equation involving its 4×4 transformation matrix becomes

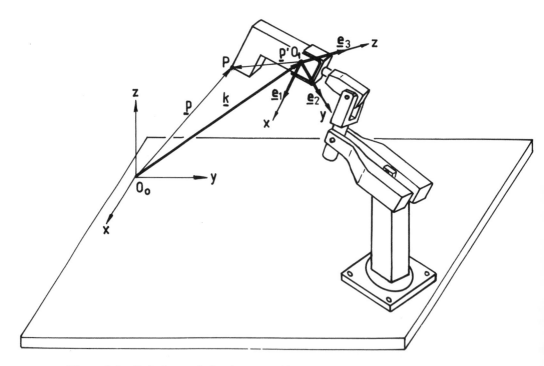

Figure 2.2   Relative and absolute coordinates. k = position vector to origin O_1 in the base coordinate system, p' = position vector to the point P in the system of origin O_1, p = position vector to the point P in the base coordinate system.

$$
\begin{bmatrix} p_x \\ p_y \\ p_z \\ 1 \end{bmatrix} = \begin{bmatrix} e_{1x} & e_{2x} & e_{3x} & k_x \\ e_{1y} & e_{2y} & e_{3y} & k_y \\ e_{1z} & e_{2z} & e_{3z} & k_z \\ 0 & 0 & 0 & 1 \end{bmatrix} * \begin{bmatrix} p'_x \\ p'_y \\ p'_z \\ 1 \end{bmatrix} \tag{2.4}
$$

Thus the homogeneous transformation matrix in Equation 2.4 to be designated by H includes rotation and translation, and can be represented as

$$
H = \left[\begin{array}{ccc:c} & R & & k \\ \hdashline 0 & 0 & 0 & 1 \end{array} \right] \tag{2.5}
$$

where submatrix R is the 3×3 rotational part of the transformation matrix and k is the 3×1 translatory part of the transformation matrix. The translatory part is a column vector pointing from the base origin O_0 to the origin O_1 of the displaced coordinate system.

One can distinguish between two types of 4×4 matrices used in robot motion calculations. The first type describes a given coordinate system relative to the base coordinate frame. The second type describes the relationship between any two coordinate systems in a chain of coordinate frames. The first type of transformation has already been implemented above and the second type will be dealt with later in Section 2.3.

2.2 MATHEMATICAL DESCRIPTION OF OBJECTS

The aim is now to show how to use homogeneous transformations for describing objects and their locations in a form understandable to the robot's computers. Objects surrounded by planar surfaces are described by $4 \times N$ matrices where N designates the number of vertices of the object. Each vertex appears as a position vector of the form

$$
[x \quad y \quad z \quad 1]^T
$$

where the superscript T stands for transpose. For the description of objects with planar surfaces there are two options:

(i) The origin of an object's coordinate system can be positioned independent of any of its features, and the orientation of its frame coordinates may also be chosen arbitrarily (see Figure 2.3(a)). The

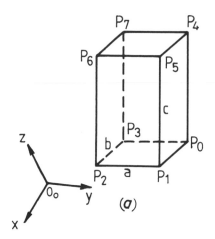 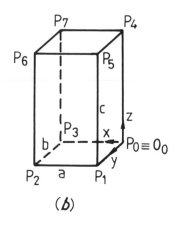

Figure 2.3 A cuboid with its eight vertices P_0–P_7. (*a*) The coordinate system of the object is independent of the features of the cuboid. (*b*) The coordinate system is fixed to the cuboid. The origin is chosen to be at the vertex P_0.

general matrix presentation of an object with N vertices is of the form

$$[\text{object}] = \begin{bmatrix} x_0 & x_1 & \cdots & x_{N-1} \\ y_0 & y_1 & \cdots & y_{N-1} \\ z_0 & z_1 & \cdots & z_{N-1} \\ 1 & 1 & \cdots & 1 \end{bmatrix} \tag{2.6}$$

(ii) The origin of an object coordinate system can be fixed to a feature of the object (see Figure 2.3(*b*)). For instance any of the vertices, center of gravity, etc., can be chosen as the origin. In our discussion one of the vertices will be adopted for convenience.

Let the object be a cuboid represented by its eight vertices P_0–P_7 in cartesian coordinates as shown in Figure 2.3(*b*). The origin of the fixed coordinate system is chosen at P_0. Such a choice makes the description of the object quite easy. The axes of the coordinate system are in alignment with the edges intersecting each other at that chosen origin. Suppose the cuboid has ends $a \times b$ units with height c units in the x, y and z directions respectively. The corresponding columns of the object's description matrix for the vertex P_0 will be of the form

$$[0 \quad 0 \quad 0 \quad 1]^{\mathrm{T}}$$

similarly for vertex P_1

$[0 \quad b \quad 0 \quad 1]^{\mathrm{T}}$

and for vertex P_2

$[a \quad b \quad 0 \quad 1]^{\mathrm{T}}$

and so forth. This leads to the description matrix of the cuboid as given in Equation 2.7, namely

$$
[\text{cuboid}] =
\begin{bmatrix}
0 & 0 & a & a & 0 & 0 & a & a \\
0 & b & b & 0 & 0 & b & b & 0 \\
0 & 0 & 0 & 0 & c & c & c & c \\
1 & 1 & 1 & 1 & 1 & 1 & 1 & 1
\end{bmatrix}
$$

(2.7)

$$
\begin{array}{cccccccc}
\uparrow & \uparrow & \uparrow & \uparrow & \uparrow & \uparrow & \uparrow & \uparrow \\
P_0 & P_1 & P_2 & P_3 & P_4 & P_5 & P_6 & P_7
\end{array}
$$

Let us now perform a translation and a rotation on the object. This procedure will be described by a 4×4 transformation matrix H as shown in Section 2.1. The relation between the new and starting positions of an object is given as

$$[\text{object}]_n = H * [\text{object}]_s \tag{2.8}$$

where n denotes the new position and s the starting position of the object, i.e. in this case the new position refers to the base coordinate frame and the starting position to the displaced coordinate system. The expanded form of Equation 2.8 becomes

$$
\begin{bmatrix}
x_{0,n} & x_{1,n} & \cdots & x_{N-1,n} \\
y_{0,n} & y_{1,n} & \cdots & y_{N-1,n} \\
z_{0,n} & z_{1,n} & \cdots & z_{N-1,n} \\
1 & 1 & \cdots & 1
\end{bmatrix}
=
\left[
\begin{array}{ccc:c}
 & R & & k \\
\hdashline
0 & 0 & 0 & 1
\end{array}
\right]
*
\begin{bmatrix}
x_{0,s} & x_{1,s} & \cdots & x_{N-1,s} \\
y_{0,s} & y_{1,s} & \cdots & y_{N-1,s} \\
z_{0,s} & z_{1,s} & \cdots & z_{N-1,s} \\
1 & 1 & \cdots & 1
\end{bmatrix}
$$

(2.9)

The matrix on the left-hand side represents the vertices of the cuboid in their new position after transformation. Matrices on the right-hand side represent the transformation matrix and the vertices of the cuboid in the starting coordinate frame respectively.

2.3   *ILLUSTRATIVE EXAMPLE 1*
DESCRIPTION OF A WEDGE BY TRANSFORMATION
MATRICES

Specification

A wedge is shown in both its initial and overturned positions in Figures 2.4(*a*)
and 2.4(*b*) respectively. A cartesian coordinate system is fixed to one of the
wedge's arbitrarily chosen vertices. In this case the origin is assigned to the
vertex at $A(0,0,0)$ shown in Figure 2.4(*a*).

Objectives

(i) Construct the description matrix W_0 of the wedge.
(ii) Describe a homogeneous transformation which translates the wedge
by 2 units along the x axis and -3 units along the z axis with zero
translation along the y axis (see Figure 2.4(*c*)).
(iii) Derive the coordinate transformation which moves the same wedge
from its original position into a new, overturned position, as shown in
Figures 2.4(*b*) and 2.4(*d*), where A→D′ and B→C′.

Solutions

(i) The description matrix W_0 of the wedge being in its initial state is derived as

$$W_0 = \begin{bmatrix} 0 & 1 & 0 & 0 & 0 & 1 \\ 0 & 0 & 1 & 0 & 1 & 1 \\ 0 & 0 & 0 & 1 & 1 & 0 \\ 1 & 1 & 1 & 1 & 1 & 1 \end{bmatrix} \tag{2.10}$$

Note that the sequence of the vertices in the matrix W_0 is arbitrary.

(ii) The shift of the wedge by 2 units along the x axis, -3 units along the z axis
and no translation along the y axis is described by the vector k in
Equation 2.11a as the fourth column in Equation 2.11b:

$$k = [2 \quad 0 \quad -3]^T \tag{2.11a}$$

There is no reorientation of the wedge thus the rotation matrix R is represented
by a unit matrix in Equation 2.11b, i.e. the direction of the coordinates (x, y, z)

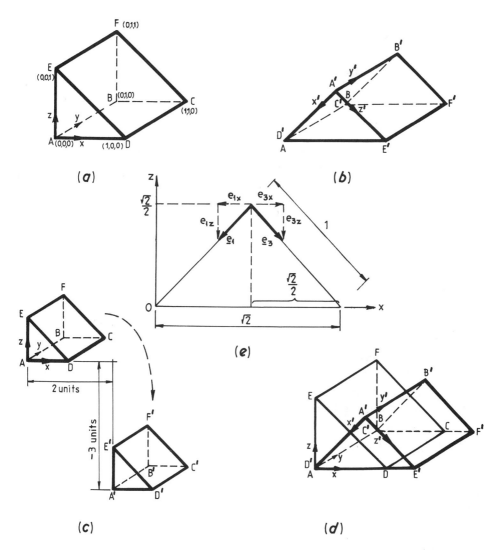

Figure 2.4 (*a*) The wedge in its cartesian coordinate system. (*b*) The new state of the wedge. (*c*) The wedge in displaced state. (*d*) The wedge in overturned state. (*e*) Support for calculation of H_2.

and (x', y', z') are identical as shown in Figure 2.4(*c*). The homogeneous transformation matrix H_1 is given in Equation 2.11b.

$$H_1 = \begin{bmatrix} 1 & 0 & 0 & 2 \\ 0 & 1 & 0 & 0 \\ 0 & 0 & 1 & -3 \\ 0 & 0 & 0 & 1 \end{bmatrix}$$

(2.11b)

The matrix W_1 describing the wedge in its translated position is given by the product of matrices H_1 and W_0 as

$$W_1 = H_1 * W_0$$

$$= \begin{bmatrix} 1 & 0 & 0 & 2 \\ 0 & 1 & 0 & 0 \\ 0 & 0 & 1 & -3 \\ 0 & 0 & 0 & 1 \end{bmatrix} * \begin{bmatrix} 0 & 1 & 0 & 0 & 0 & 1 \\ 0 & 0 & 1 & 0 & 1 & 1 \\ 0 & 0 & 0 & 1 & 1 & 0 \\ 1 & 1 & 1 & 1 & 1 & 1 \end{bmatrix}$$

$$= \begin{bmatrix} 2 & 3 & 2 & 2 & 2 & 3 \\ 0 & 0 & 1 & 0 & 1 & 1 \\ -3 & -3 & -3 & -2 & -2 & -3 \\ 1 & 1 & 1 & 1 & 1 & 1 \end{bmatrix} \qquad (2.12)$$

(iii) When the wedge is turned over from its original position as shown in Figure 2.4(a) into another state as shown in Figures 2.4(b) and 2.4(d), the homogeneous transformation is constructed as given below. By looking at this new state of the wedge we can see how its position and orientation have been changed. The coordinates of the vertex of the new origin at A' can be calculated from Figure 2.4(e) as

$$x_0 = \frac{\sqrt{2}}{2}$$

$$y_0 = 0$$

$$z_0 = \frac{\sqrt{2}}{2}$$

The new (x, y, z) axes, corresponding to the unit vector triad (e_1, e_2, e_3), are as follows

$$e_1 = \begin{bmatrix} e_{1x} \\ e_{1y} \\ e_{1z} \end{bmatrix} = \begin{bmatrix} -\sqrt{2}/2 \\ 0 \\ -\sqrt{2}/2 \end{bmatrix}$$

$$e_2 = \begin{bmatrix} e_{2x} \\ e_{2y} \\ e_{2z} \end{bmatrix} = \begin{bmatrix} 0 \\ 1 \\ 0 \end{bmatrix}$$

$$e_3 = \begin{bmatrix} e_{3x} \\ e_{3y} \\ e_{3z} \end{bmatrix} = \begin{bmatrix} \sqrt{2}/2 \\ 0 \\ -\sqrt{2}/2 \end{bmatrix}$$

Thus the transformation matrix H_2 is given as

$$H_2 = \begin{bmatrix} -\sqrt{2}/2 & 0 & \sqrt{2}/2 & \sqrt{2}/2 \\ 0 & 1 & 0 & 0 \\ -\sqrt{2}/2 & 0 & -\sqrt{2}/2 & \sqrt{2}/2 \\ 0 & 0 & 0 & 1 \end{bmatrix} \tag{2.13}$$

The description matrix W_2 of the wedge in its overturned position shown in Figure 2.4(d) becomes

$$W_2 = H_2 * W_0$$

$$= \begin{bmatrix} -\sqrt{2}/2 & 0 & \sqrt{2}/2 & \sqrt{2}/2 \\ 0 & 1 & 0 & 0 \\ -\sqrt{2}/2 & 0 & -\sqrt{2}/2 & \sqrt{2}/2 \\ 0 & 0 & 0 & 1 \end{bmatrix} * \begin{bmatrix} 0 & 1 & 0 & 0 & 0 & 1 \\ 0 & 0 & 1 & 0 & 1 & 1 \\ 0 & 0 & 0 & 1 & 1 & 0 \\ 1 & 1 & 1 & 1 & 1 & 1 \end{bmatrix}$$

$$W_2 = \begin{bmatrix} \sqrt{2}/2 & 0 & \sqrt{2}/2 & \sqrt{2} & \sqrt{2} & 0 \\ 0 & 0 & 1 & 0 & 1 & 1 \\ \sqrt{2}/2 & 0 & \sqrt{2}/2 & 0 & 0 & 0 \\ 1 & 1 & 1 & 1 & 1 & 1 \end{bmatrix} \tag{2.14}$$

Equation 2.14 provides the new coordinates of the wedge's vertices in its overturned position. This may also be determined intuitively by calculating the vertices of the translated wedge and then by forming the wedge description matrix.

2.4   RELATIVE TRANSFORMATIONS IN THE ROBOT WORKSPACE

From a mathematical point of view the robotic interpretation of an assembly can be considered as a sequence of homogeneous coordinate transformations.

Targets of the motion are usually given in terms of spatial locations. The features of parts to be manipulated are defined relative to features of other assembly parts.

For the purpose of describing an assembly of parts in robot programming, normally a relative coordinate transformation is used. A great advantage of using a relative transformation is that if later on for some reason any modification is required in the positions, orientations and/or dimensions to any composite part of the assembly, or if further spatial adjustment to the whole assembly is necessary, only those segments of the program where the modifications have been applied need to be altered. The rest of the computer program remains unaffected. As an example, let us consider the assembly shown in Figure 2.5, which will be dealt with numerically in Illustrative example 2.

The position of the target point P_t in the base coordinate system of origin O_0 is defined by the geometry of the assembled parts, their relative position and orientation, and by the absolute position of the origin O_1. Suppose we want to move the robot manipulator's gripper or tool to the target point P_t on the part C. By knowing all dimensions of the objects A, B and C as well as their locations and orientations in space, we can determine the position of the point P_t, i.e. we

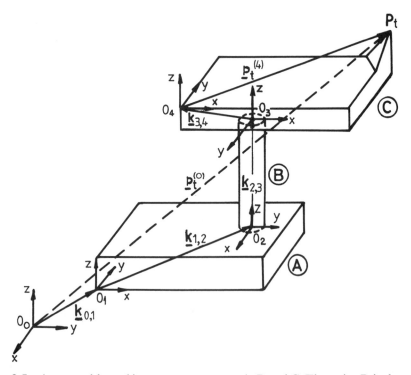

Figure 2.5   An assembly and its component parts A, B and C. The point P_t is the target to be reached by the manipulator.

can obtain its coordinates by using homogeneous transformations.

In order to demonstrate this process we approach the target point P_t step by step from the origin O_0. Before doing this we assign a separate coordinate system to each part of the assembly. For data input we shall use the coordinates of P_t in the coordinate system of origin O_4 which are designated by

$$x_t^{(4)}, \quad y_t^{(4)}, \quad z_t^{(4)}$$

The aim is now to express the position of the target P_t in the base coordinate system, that is we want to set an expression for the vector $p_t^{(0)}$ whose coordinates are

$$x_t^{(0)}, \quad y_t^{(0)}, \quad z_t^{(0)}$$

The transformation matrix $H_{i,n}$ describing the position and orientation of the nth coordinate system relative to the ith one can be formulated as

$$H_{i,n} = \prod_{i}^{n-1} H_{i,i+1} = H_{i,i+1} * H_{i+1,i+2} * \cdots * H_{n-1,n} \tag{2.15}$$

where $i = 0, 1, \ldots, n-1$ which may start at any number for $i < n$ and

$$H_{i,i+1} = \left[\begin{array}{ccc:c} & R_{i,i+1} & & k_{i,i+1} \\ \hdashline 0 & 0 & 0 & 1 \end{array} \right] \tag{2.16}$$

The resultant transformation matrix $H_{i,n}$ in Equation 2.15 describing the state of the nth coordinate system relative to any ith component's coordinate frame is the product of sequentially constituted transformations for $i < n$. Thus the coordinates of the target point P_t in the nth coordinate system relative to any ith component's coordinate frame can be expressed as in Equations 2.17a and 2.17b respectively.

$$p_t^{(i)} = H_{i,n} * p_t^{(n)} \tag{2.17a}$$

or in an expanded form:

$$\begin{bmatrix} x_t^{(i)} \\ y_t^{(i)} \\ z_t^{(i)} \\ 1 \end{bmatrix} = H_{i,n} * \begin{bmatrix} x_t^{(n)} \\ y_t^{(n)} \\ z_t^{(n)} \\ 1 \end{bmatrix} \tag{2.17b}$$

For instance in the case shown in Figure 2.5, the vector $p_t^{(0)}$ and its coordinates to the target point P_t, for $i=0$ and $n=4$, are given in the base coordinate system as

$$p_t^{(0)} = H_{0,4} * p_t^{(4)} \tag{2.18a}$$

and in expanded form

$$\begin{bmatrix} x_t^{(0)} \\ y_t^{(0)} \\ z_t^{(0)} \\ 1 \end{bmatrix} = H_{0,1} * H_{1,2} * H_{2,3} * H_{3,4} * \begin{bmatrix} x_t^{(4)} \\ y_t^{(4)} \\ z_t^{(4)} \\ 1 \end{bmatrix} \tag{2.18b}$$

2.5   ILLUSTRATIVE EXAMPLE 2
DERIVATION OF TRANSFORMATION MATRICES ALONG AN ASSEMBLY

Specification

An assembled workpiece is shown in Figure 2.5. The coordinates of the base system at origin O_0 are (0, 0, 0) and those of origin O_1 relative to the base system are $(-5, 10, 10)$. The coordinates of the target point P_t in the system of origin O_4 are given by

$$x_t^{(4)} = 5 \quad y_t^{(4)} = 3 \quad z_t^{(4)} = 6$$

The heights of parts A, B and C are 2, 10 and 1 units respectively. All other dimensions of the parts required in the calculation are obtainable from Figure 2.6.

Objectives

(i) Redraw Figure 2.5 and show all numerical values of the coordinates along the parts of the workpiece.

(ii) Calculate the coordinates of the target point P_t in the base system according to Equations 2.17 by deriving the homogeneous transformation matrices leading from origin O_0 to origin O_4. Check the result by simple addition of the dimensions.

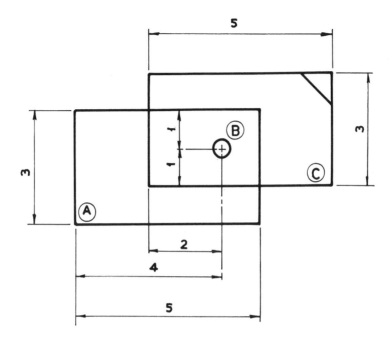

Figure 2.6 Top view of the assembly in Figure 2.5.

Solutions

(i) The numerical dimensions of the parts of the assembly are given in Figure 2.7.

(ii) The relative homogeneous transformation matrices between O_0 and O_4 are calculated and given in a sequential order as

$$H_{0,1} = \begin{bmatrix} 0 & -1 & 0 & -5 \\ 1 & 0 & 0 & 10 \\ 0 & 0 & 1 & 10 \\ 0 & 0 & 0 & 1 \end{bmatrix}$$

$$H_{1,2} = \begin{bmatrix} 0 & 1 & 0 & 4 \\ -1 & 0 & 0 & 2 \\ 0 & 0 & 1 & 2 \\ 0 & 0 & 0 & 1 \end{bmatrix}$$

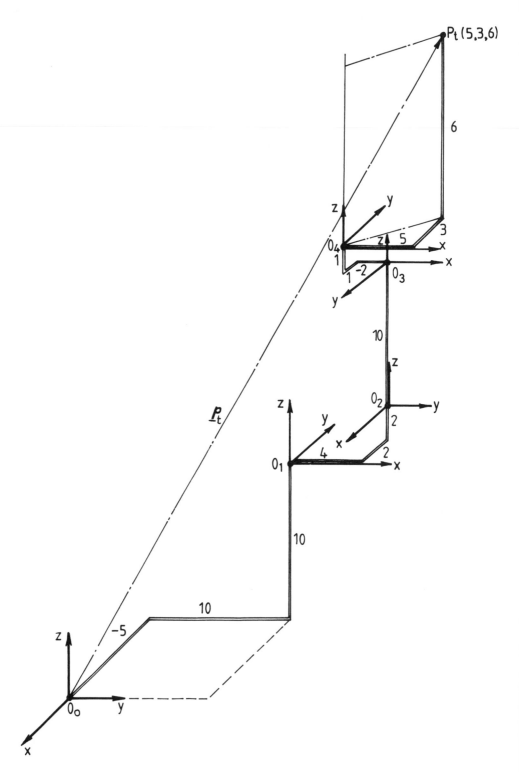

Figure 2.7 Calculated values of the coordinates along the assembly in Figure 2.5.

$$H_{2,3} = \begin{bmatrix} 0 & 1 & 0 & 0 \\ 1 & 0 & 0 & 0 \\ 0 & 0 & 1 & 10 \\ 0 & 0 & 0 & 1 \end{bmatrix}$$

$$H_{3,4} = \begin{bmatrix} 1 & 0 & 0 & -2 \\ 0 & -1 & 0 & 1 \\ 0 & 0 & 1 & 1 \\ 0 & 0 & 0 & 1 \end{bmatrix} \tag{2.19}$$

Thus the overall transformation matrix $H_{0,4}$ is the product of the four transformation matrices in Equations 2.19 and is given as

$$H_{0,4} = \begin{bmatrix} 0 & -1 & 0 & -6 \\ 1 & 0 & 0 & 12 \\ 0 & 0 & 1 & 23 \\ 0 & 0 & 0 & 1 \end{bmatrix} \tag{2.20}$$

Finally the numerical values of the target point's coordinates in the base system are obtained from Equation 2.17.

$$\begin{bmatrix} x_t^{(0)} \\ y_t^{(0)} \\ z_t^{(0)} \\ 1 \end{bmatrix} = \begin{bmatrix} 0 & -1 & 0 & -6 \\ 1 & 0 & 0 & 12 \\ 0 & 0 & 1 & 23 \\ 0 & 0 & 0 & 1 \end{bmatrix} * \begin{bmatrix} 5 \\ 3 \\ 6 \\ 1 \end{bmatrix} = \begin{bmatrix} -9 \\ 17 \\ 29 \\ 1 \end{bmatrix} \tag{2.21}$$

that is $x_t^{(0)} = -9$, $y_t^{(0)} = 17$, $z_t^{(0)} = 29$.

The superscript $^{(0)}$ in round brackets refers to the origin of the base coordinate system.

Note that any dimensional change anywhere along the assembly necessitates only one transformation related to the corresponding element and there is no further effect on the other assembly parts.

2.6 TRANSFORMATIONS ALONG THE KINEMATIC CHAIN

An anthropomorphic robot arm, otherwise called a serial link manipulator or robot manipulator, consists of a sequence of links connected by activated joints [2.2]. The robot may also have partly or wholly prismatic joints, as outlined in the classification of robots in Section 1.5. Therefore in the following we will discuss both revolute and prismatic types.

A robot manipulator of n degrees of mobility (see Section 1.8) is considered to have n joints connected by n links. The assembly of sequential links and joints makes up a kinematic chain. The function of the links is to maintain a fixed relationship between the manipulator joints at each end of the link.

In order to describe the relationship between the elements of the kinematic chain, we will assign a coordinate system to each joint. Any link can be characterized by two quantities, i.e.

> (1) The distance a_i along the common normal (CN) between the diverging joint axes at both ends of the link.
> (2) The angle α_i between the diverging joint axes perpendicular to a_i.

a_i is usually called the link length and α_i is the link twist angle (see Figure 2.8). In the case of no diversion, i.e. $\alpha_i = 0$, there are an infinite number of common normals.

In the case of a serial link manipulator two links are connected at each joint axis (see Figure 2.9). Every joint axis has two significant common normals, one for each joining link. The relative position of these two connected links is given by b_i, i.e. the distance between the two common normals along the particular joint axis. The ϑ_i is the joint angle between these two normals, measured in a

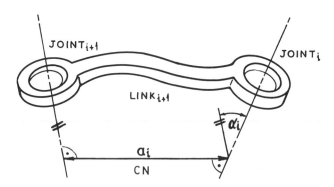

Figure 2.8 Robot link and its parameters; a_i = the link length along the CN, α_i = the link angle between the joint axes.

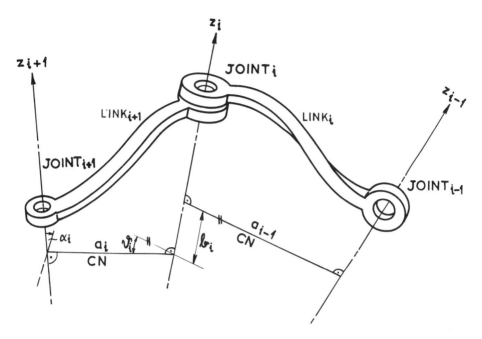

Figure 2.9 Robot joints and their parameters; b_i = the joint distance, ϑ_i = the joint angle.

perpendicular plane to the joint axis z_i. So the distance b_i and the joint angle ϑ_i are the variables of that particular joint.

In a robot's structure we distinguish between revolute and prismatic joints. In the first case the link turns (twists, bends) about its axis and α_i and/or ϑ_i are the joint variables. In the latter case the link moves along its axis and a_i and /or b_i are the variables (see Figures 2.8, 2.9 and 2.10).

For assigning coordinate systems to the members of the kinematic chains, the Denavit–Hartenberg convention has been commonly adopted in manipulator geometry [2.3] [2.4]. Following this convention, a coordinate system is assigned to each joint of the robot manipulator so that a defined sequence of at most two rotations and two translations is required to bring the ith coordinate system into coincidence with the $(i+1)$th system [2.5]. This procedure is demonstrated for joint_i and joint_{i+1} as shown in Figure 2.11.

In order to demonstrate the coincidence procedure in Figure 2.11, consider the individual steps as follows:

    (i) Turn about axis z_i by an angle ϑ_i, $\boldsymbol{H}\,(\vartheta_i)$.
    (ii) Move along axis z_i by a distance b_i, $\boldsymbol{H}\,(b_i)$.
    (iii) Move along axis x_{i+1} by a distance a_i, $\boldsymbol{H}\,(a_i)$.
    (iv) Turn about axis x_{i+1} by an angle α_i, $\boldsymbol{H}\,(\alpha_i)$.

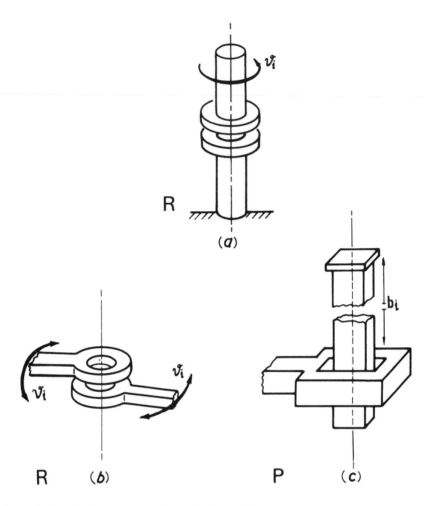

Figure 2.10 Typical types of manipulator joints; R = the revolute (twist or bend) joints shown in (a) and (b), P = the prismatic joint shown in (c).

Let p_i and p_{i+1} represent a point P in the ith and $(i+1)$th coordinate systems as follows

$$p_i = \begin{bmatrix} x_p \\ y_p \\ z_p \\ 1 \end{bmatrix}_i \qquad (2.22)$$

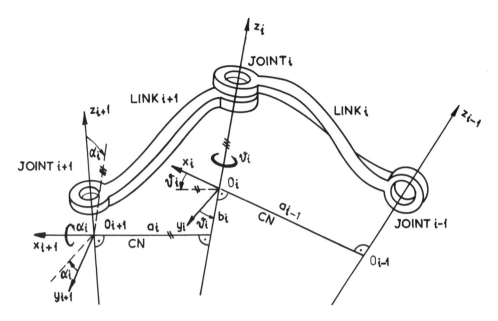

Figure 2.11 Joint coordinate system of a kinematic chain. To each joint a coordinate set is assigned. The parameters are defined in accordance with the Denavit–Hartenberg convention.

$$p_{i+1} = \begin{bmatrix} x_p \\ y_p \\ z_p \\ 1 \end{bmatrix}_{i+1} \tag{2.23}$$

The relationship between these two joints according to Equation 2.17 is

$$p_i = H_{i,i+1} * p_{i+1} \tag{2.24}$$

where matrix $H_{i,i+1}$ is made up of four coordinate transformations outlined above under (i)–(iv). We now consider each step of the coincidence procedure in subsequent order shown in resolved states in Figure 2.12 (see Appendix 2.A).

(i) The transformation matrix for the rotation ϑ_i about the z_i axis corresponding to Figure 2.12(a) is

$$H(\vartheta_i) = \begin{bmatrix} \cos \vartheta_i & -\sin \vartheta_i & 0 & 0 \\ \sin \vartheta_i & \cos \vartheta_i & 0 & 0 \\ 0 & 0 & 1 & 0 \\ 0 & 0 & 0 & 1 \end{bmatrix} \tag{2.25}$$

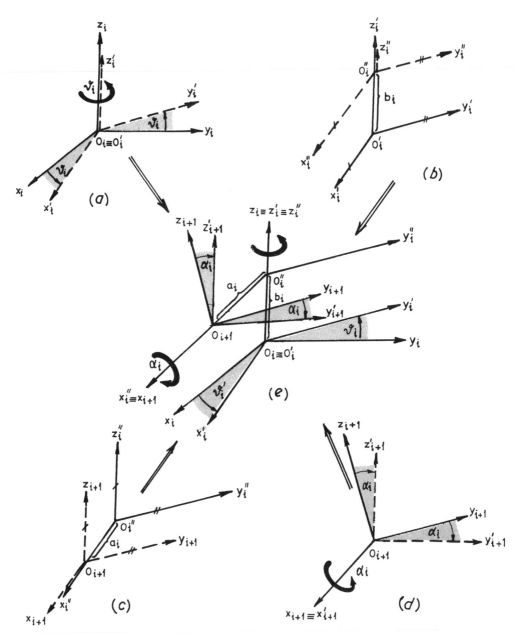

Figure 2.12 Homogeneous coordinate transformation. (*a*) Rotation ϑ_i about the z_i axis. (*b*) Displacement along the z_i axis. (*c*) Displacement along the x_{i+1} axis. (*d*) Rotation α_i about the x_{i+1} axis. (*e*) Resultant transformation between joint$_i$ and joint$_{i+1}$.

(ii) The transformation matrix for the translation b_i along the axis z_i corresponding to Figure 2.12(b) is

$$H(b_i) = \begin{bmatrix} 1 & 0 & 0 & 0 \\ 0 & 1 & 0 & 0 \\ 0 & 0 & 1 & b_i \\ 0 & 0 & 0 & 1 \end{bmatrix} \tag{2.26}$$

(iii) The transformation matrix for the translation a_i along the axis x_{i+1} corresponding to Figure 2.12(c) becomes

$$H(a_i) = \begin{bmatrix} 1 & 0 & 0 & a_i \\ 0 & 1 & 0 & 0 \\ 0 & 0 & 1 & 0 \\ 0 & 0 & 0 & 1 \end{bmatrix} \tag{2.27}$$

(iv) The transformation matrix for the rotation α_i about the x_{i+1} axis corresponding to Figure 2.12(d) is

$$H(\alpha_i) = \begin{bmatrix} 1 & 0 & 0 & 0 \\ 0 & \cos \alpha_i & -\sin \alpha_i & 0 \\ 0 & \sin \alpha_i & \cos \alpha_i & 0 \\ 0 & 0 & 0 & 1 \end{bmatrix} \tag{2.28}$$

The resultant transformation matrix between two adjacent joints corresponds to Figure 2.12(e) and according to Equation (2.16) is given by

$$H_{i,i+1} = H(\vartheta_i) * H(b_i) * H(a_i) * H(\alpha_i)$$

or in expanded form as

$$H_{i,i+1} = \begin{bmatrix} \cos \vartheta_i & -\sin \vartheta_i \cos \alpha_i & \sin \vartheta_i \sin \alpha_i & a_i \cos \vartheta_i \\ \sin \vartheta_i & \cos \vartheta_i \cos \alpha_i & -\cos \vartheta_i \sin \alpha_i & a_i \sin \vartheta_i \\ 0 & \sin \alpha_i & \cos \alpha_i & b_i \\ 0 & 0 & 0 & 1 \end{bmatrix} \tag{2.29}$$

The matrix in Equation (2.29) is the overall transformation matrix in the most general form between two subsequent joints, i.e. between the ith and $(i+1)$th joints. The rotation occurs about a single axis, that is ϑ_i about the $z_i' \equiv z_i$ axis and α_i about the $x_i'' = x_{i+1}$ axis. The fourth column of the $H_{i,i+1}$ matrix is the position vector k between the two origins O_i and O_{i+1}. After four steps the coincidence has been obtained shown in Figure 2.12(e).

2.7   DESCRIPTION OF MANIPULATOR JOINTS

We first consider the revolute (R-type) joints where, as stated in Section 2.6, either ϑ_i or α_i is the joint variable. A single joint is supposed to rotate only about one axis. As shown in Figure 2.13, the z_i axis is always the axis of rotation of the joint$_i$ and the x_i axis is in alignment with the common normal CN of the axes z_{i-1} and z_i. The origin O_i of the ith coordinate system is set to be at the intersection of the axis of joint$_i$, with the common normal being between the axes of joint$_{i-1}$ and joint$_i$. From a geometrical point of view the robot manipulator has two exceptions, i.e. when the z axes of adjacent joints are either parallel or intersect.

In the case of parallel axes (see Figure 2.14), the origin O_i is set on the z_i axis to make the distance b_i zero. The z_{i-1} axis of the joint$_{i-1}$ will be aligned with the z_i axis of joint$_i$. There is an infinite number of common normals and the x_i axis will be aligned with the common normal directed from the ith origin O_i.

In the case of intersecting axes the origin O_i of the ith coordinate system will be at the intersection point of the z_{i-1} and z_i axes. The direction of the axis x_i is either parallel or antiparallel to the direction of the vector cross product $z_{i-1} \times z_i$ (see Figure 2.15). This condition has been satisfied in the case of the x_i axis being directed along the common normal between joint$_{i-1}$ and joint$_i$. We

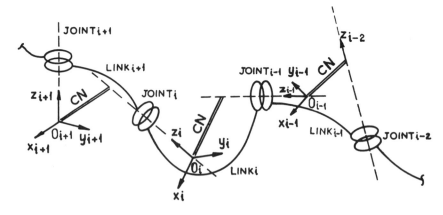

Figure 2.13   Relationship of the joint coordinate systems assigned to adjacent revolute joints.

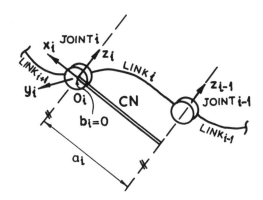

Figure 2.14 Parallel joint axes of adjacent revolute joints. The joint distance b_i becomes zero, the z_{i-1} and z_i axes are parallel ($\alpha_i = 0$) and x_i axis is in alignment with the common normal.

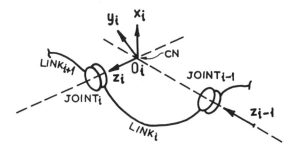

Figure 2.15 Intersecting joint axes of adjacent revolute joints. The origin O_i of the ith coordinate system is at the point of intersection of the z_i and z_{i-1} axes.

note that for this case the CN is a point and the conventional robot manipulator usually is of this kind of structure.

Now let us consider the prismatic joint (P-type), where either a_i or b_i is the joint variable. If it translates in the direction of a joint's z axis, then b_i is the prismatic joint variable. The origin O_i of the coordinate system of a prismatic joint is in the intersection point of the z_i axis and the common normal of the axes z_{i-1} and z_i shown in Figure 2.16.

For the parallel and intersecting prismatic joint axes the same rules apply as for revolute joints except that the common normal of parallel axes pointing to the ith origin starts from the $(i-1)$th origin (see Figure 2.17). For adjacent parallel prismatic joints the position of the ith origin on the z axis is arbitrary. For a prismatic joint we define its zero position for $b_i = 0$ when x_{i-1} and x_i are parallel and have the same direction.

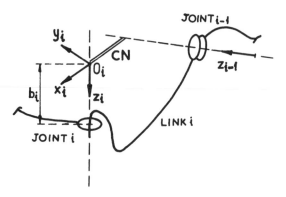

Figure 2.16   Prismatic joint configuration.

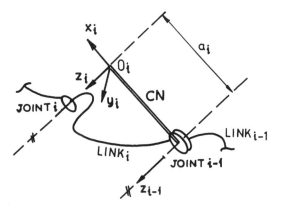

Figure 2.17   Adjacent prismatic and revolute joints with parallel axes. The position of O_i along the axis of the ith joint is arbitrary.

2.8   *ILLUSTRATIVE EXAMPLE 3*
ASSIGNMENT OF COORDINATE SYSTEMS TO
ROBOT JOINTS AND DERIVATION OF THE
TRANSFORMATION MATRICES

Specification

A jointed linkage mechanism is shown in Figure 2.18. The mechanism is of the
RRRRP-type, i.e. it has four revolute joints J1–J4, two bending and two swivel
types, and one prismatic joint J5.

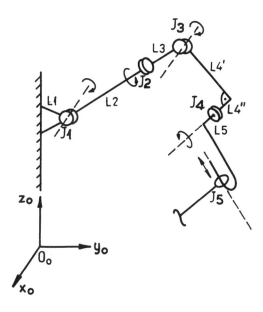

Figure 2.18   Linkage-joint mechanism of RRRRP-type robot.

Objectives

(i) Assign a coordinate system to each joint of the mechanism.
(ii) Derive the transformation matrices between each pair of joints.

Solutions

(i) The assignment of a coordinate system to each joint is processed in the following five steps as shown in Figure 2.19.

(a) Joint J1 is a revolute (bend) joint. The joint variable is the angle q_1. The z_1 axis of the coordinate system lies in the joint axis. The x_1 axis lies in $\overline{\text{AO}_1}$ which is the common normal (CN) to axes z_0 and z_1. The y_1 axis is perpendicular to z_1 and x_1 which completes the right-hand rotating coordinate system, i.e.

$$x_1 = z_0 \rangle\langle z_1 \text{ and } y_1 = z_1 \rangle\langle x_1$$

(b) Joint J2 is a revolute (swivel) joint. The joint variable is the angle q_2. The z_2 axis aligns with the axis of joint J2. The z_2 axis intersects z_1, i.e.

$$x_2 = z_1 \rangle\langle z_2 \text{ and } y_2 = z_2 \rangle\langle x_2$$

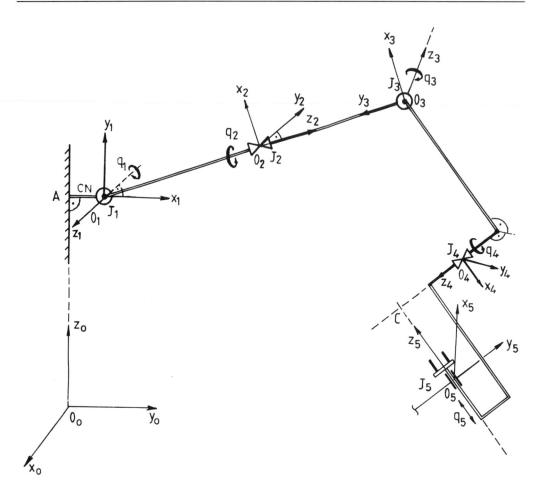

Figure 2.19 Assignment of coordinate systems to the joints.

(c) Joint J3 is a revolute (bend) joint. The joint variable is the angle q_3. The axis z_3 lies in the joint axis. The z_3 axis intersects z_2, therefore

$$x_3 = z_2 \backslash z_3 \text{ and } y_3 = z_3 \backslash x_3$$

(d) Joint J4 is a revolute (swivel) joint. The joint variable is q_4. The z_4 axis lies in the axis of joint J4. The origin O_4 is at the point where the common normal of z_3 and z_4 intersects z_4. The x_4 axis lies in the common normal of z_3 and z_4, i.e.

$$x_4 = z_3 \backslash z_4 \text{ and } y_4 = z_4 \backslash x_4$$

(e) Joint J5 is a prismatic joint. The joint variable is the offset q_5. The z_5 axis aligns with the axis of joint J5 and intersects z_4, i.e.

$$x_5 = z_4 \times z_5 \text{ and } y_5 = z_5 \times x_5$$

(ii) Transformation matrices between adjacent joints are derived from the reference coordinate system of the robot shown in Figure 2.20. The reference state represents a special joint-link configuration of the robot manipulator at

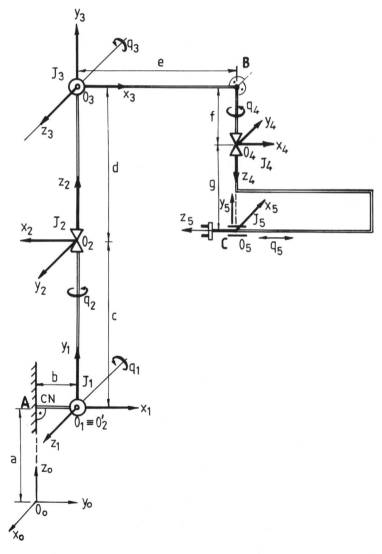

Figure 2.20 The reference state of the manipulator and the assigned reference coordinate systems of the joints.

hand for which each joint variable is zero. According to the chosen reference configuration, the joint variables have special values as follows

$q_0=0$ if $x_1 \perp x_0$, i.e. $\vartheta_0=90°$ x_0 and x_1 are perpendicular
$q_1=0$ if $x_2 \parallel x_1$, i.e. $\vartheta_1=q_1-180°$ x_1 and x_2 are antiparallel
$q_2=0$ if $x_3 \parallel x_2$, i.e. $\vartheta_2=q_2-180°$ x_2 and x_3 are antiparallel
$q_3=0$ if $x_4 \parallel x_3$, i.e. $\vartheta_3=q_3$ x_3 and x_4 are parallel
$q_4=0$ if $x_5 \perp x_4$, i.e. $\vartheta_4=90°$ x_4 and x_5 are perpendicular
$q_5=0$ can be chosen arbitrarily x_5 is perpendicular to z_4 and z_5
within its mechanical limits

The link distances and the length of links of the robot mechanism are denoted as shown in Figure 2.20.

$$\overline{O_0A}=a,\ \overline{AO_1}=b,\ \overline{O_1O_2}=c,\ \overline{O_2O_3}=d,\ \overline{O_3B}=e,\ \overline{BO_4}=f,\ \overline{O_4O_5}=g$$

The link parameters and the individual joint transformation matrices are derived as described in Sections 2.5 and 2.6. To compute the explicit values of the members of the transformation matrices, use Equation 2.29 and consult both Figures 2.11 and 2.20 simultaneously. Let us now substitute the corresponding variables into Equation 2.29; the transformation matrices for the coordinate frames of the joints are obtained in sequence as follows.

(a) Base coordinate frame:

$$\vartheta_0=90°,\quad b_0=a,\quad a_0=b,\quad \alpha_0=90°$$

$$H_{0,1}=\begin{bmatrix} 0 & 0 & 1 & 0 \\ 1 & 0 & 0 & b \\ 0 & 1 & 0 & a \\ 0 & 0 & 0 & 1 \end{bmatrix} \tag{2.30}$$

(b) Joint J1:

$$\vartheta_1=q_1-180°,\quad b_1=0,\quad a_1=0,\quad \alpha_1=90°$$

$$H_{1,2}=\begin{bmatrix} -\cos q_1 & 0 & -\sin q_1 & 0 \\ -\sin q_1 & 0 & \cos q_1 & 0 \\ 0 & 1 & 0 & 0 \\ 0 & 0 & 0 & 1 \end{bmatrix} \tag{2.31}$$

The center of rotation q_2, i.e. the position of the origin O_2 along section $\overline{O_1 O_3}$ is arbitrarily chosen. We consider it to be identical with O_1 and is denoted as O_2' in Figure 2.20.

(c) Joint J2:

$$\vartheta_2 = q_2 - 180°, \quad b_2 = c + d, \quad a_2 = 0, \quad \alpha_2 = 90°$$

$$H_{2,3} = \begin{bmatrix} -\cos q_2 & 0 & -\sin q_2 & 0 \\ -\sin q_2 & 0 & \cos q_2 & 0 \\ 0 & 1 & 0 & c+d \\ 0 & 0 & 0 & 1 \end{bmatrix} \tag{2.32}$$

(d) Joint J3:

$$\vartheta_3 = q_3, \quad b_3 = 0, \quad a_3 = e, \quad \alpha_3 = 90°$$

$$H_{3,4} = \begin{bmatrix} \cos q_3 & 0 & \sin q_3 & e\cos q_3 \\ \sin q_3 & 0 & -\cos q_3 & e\sin q_3 \\ 0 & 1 & 0 & 0 \\ 0 & 0 & 0 & 1 \end{bmatrix} \tag{2.33}$$

(e) Joint J4:

$$\vartheta_4 = q_4, \quad b_4 = f + g, \quad a_4 = 0, \quad \alpha_4 = 90°$$

$$H_{4,5} = \begin{bmatrix} \cos q_4 & 0 & \sin q_4 & 0 \\ \sin q_4 & 0 & -\cos q_4 & 0 \\ 0 & 1 & 0 & f+g \\ 0 & 0 & 0 & 1 \end{bmatrix} \tag{2.34}$$

(f) Joint J5:
The position of O_5 along z_5 in the direction of the prismatic link is arbitrary but in our case is placed at the intersection point of axes z_4 and z_5. Let us note that the common normal to z_4 and z_5 is of zero length.

By performing the transformations from joint J1 to J5, the resultant transformation matrix $H_{0,5}$ equals

$$H_{0,5} = H_{0,1} * H_{1,2} * H_{2,3} * H_{3,4} * H_{4,5}$$

The position coordinates of the last joint are

$$x_5 = 0, \quad y_5 = b + e, \quad z_5 = a + c + d - f - g$$

which can be seen in Figure 2.20.

2.9   APPENDIX

Appendix 2.A Trigonometric formulas

$$\sin^2 A + \cos^2 A = 1$$

$$\sin 2A = 2 \sin A \cos A$$
$$\cos 2A = \cos^2 A - \sin^2 A = 2 \cos^2 A - 1 = 1 - 2 \sin^2 A$$

$$\sin(A + B) = \sin A \cos B + \cos A \sin B \tag{2.A.1}$$

$$\sin(A - B) = \sin A \cos B - \cos A \sin B \tag{2.A.2}$$

$$\cos(A + B) = \cos A \cos B - \sin A \sin B \tag{2.A.3}$$

$$\cos(A - B) = \cos A \cos B + \sin A \sin B \tag{2.A.4}$$

2.10   REFERENCES

[2.1]   Paul, R. The mathematics of computer controlled manipulators, *Proc. of 1977 JACC*, June 1977, pp. 124–31.
[2.2]   Paul, R., Shimano, B. and Mayer, G. E. Kinematic control equations for simple manipulators, *IEEE Tr. and Systems, Man and Cybernetics*, Vol. SMC-11, No. 6, June 1981.
[2.3]   Hartenberg, R. S. and Denavit, J. A kinematic notation for lower pair mechanisms based on matrices, *J. Appl. Mech.*, June 1955, pp. 215–21.
[2.4]   Myers, D. and Gordon, D. Kinematic equations for industrial manipulators, *The Industrial Robot*, September 1982, pp. 162–5.
[2.5]   Siegler, A. Kinematics and microcomputer control of a 6 degrees of freedom manipulator, *Research Report*, Cambridge University Engineering Department, CUED-CMS 185/1979.

3

COORDINATE SYSTEMS IN ROBOT APPLICATIONS

3.1   EULER ANGLES FOR SPECIFYING ORIENTATION

As discussed in Section 2.1, the orientation of a coordinate system can be specified by a rotation matrix R which is an upper left orthogonal 3×3 submatrix of the 4×4 homogeneous transformation matrix given in Equations 2.1–2.5. In robot applications a frequent task is to specify the relative orientation of an object in terms of Euler angles. The relative orientation of a coordinate system with respect to another one can be specified by a sequence of rotations about the x, y, z axes of the reference coordinate system. The order in which rotations are performed is significant. These rotations are given in terms of three angular variables, usually referred to as Euler angles [3.1]. There is no convention as to how to choose them. The choice depends on the application in question. According to our choice, Euler angles describe an arbitrary orientation in terms of rotation about the three axes. Under this interpretation quite a number of conventional rotation sets are available for use. The most frequently used rotation sequences in industrial robot applications are:

    (i) roll–yaw–roll.
    (ii) roll–pitch–yaw.
    (iii) roll–pitch–roll.

These three wrist rotation configurations will be explained by the motion of an airplane and then we will describe them in mathematical detail. The motion of an airplane can be described by the translation of the center of gravity and by the rotations about this point. A set of cartesian axes, the principal axes, is affixed to the airplane shown in Figure 3.1(*a*). The principal axes are labeled z, y and x and the angular velocities around these axes are labeled ϕ for roll, θ for pitch, and ψ for yaw respectively.

There are three primary surfaces used to control the attitude of the airplane. These are the elevators, rudder and ailerons shown in Figure 3.1(*b*), (*c*) and (*d*) respectively, together with the principal rotations produced by each. It is important to keep in mind that these motions are interdependent and operate in a similar manner as in the wrist of the robot. The figures are self-explanatory and it is easy to understand the principal rotations of a robot wrist.

(1) Roll–yaw–roll geometry

In the roll–yaw–roll convention, roll stands for rotation about the z axis, yaw about the new x axis designated by x' and roll about the new $z(\equiv z')$ axis designated by $z''(\equiv z''')$. The roll–yaw–roll robot geometry with the corresponding Euler angles is illustrated in Figure 3.2.

    The matrix of the overall rotation $R_{RYR}(\alpha, \beta, \gamma)$ is clearly the product of the

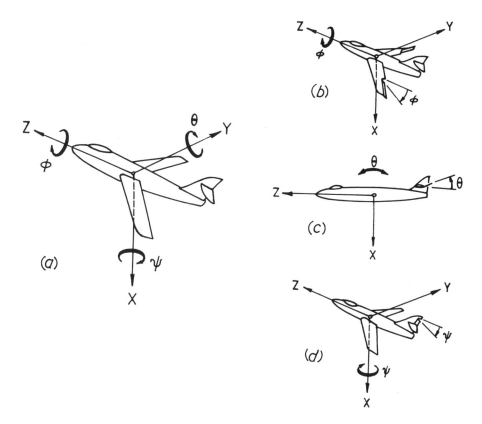

Figure 3.1 Airplane control surfaces. (*a*) The *z*, *y*, *x*, the principal axes of the airplane. (*b*) Aileron deflection produces roll angle ϕ. (*c*) Elevator deflection produces pitch angle θ. (*d*) Rudder deflection produces yaw angle ψ.

three single axis rotation matrices corresponding to the individual axis rotations α, β and γ:

$$R_{\text{RYR}}(\alpha,\ \beta,\ \gamma) = R_z(\alpha) * R_{x'}(\beta) * R_{z''}(\gamma) \tag{3.1}$$

where

α = rotation about the *z* axis – roll
β = rotation about the *x'* axis – yaw
γ = rotation about the *z''* axis – roll

The rotation matrices corresponding to the single axis rotations are as follows:

$$R_z(\alpha) = \begin{bmatrix} \cos \alpha & -\sin \alpha & 0 \\ \sin \alpha & \cos \alpha & 0 \\ 0 & 0 & 1 \end{bmatrix}_{\text{roll}} \tag{3.2}$$

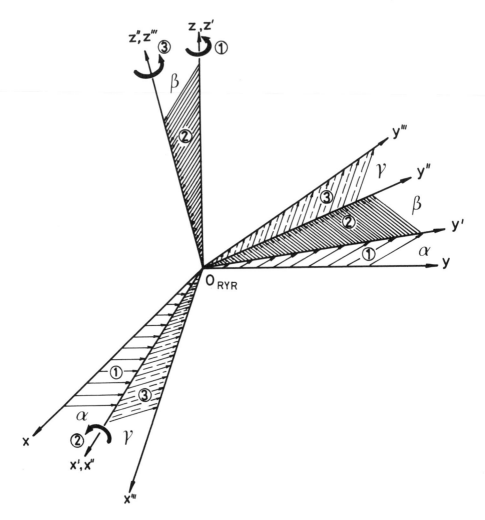

Figure 3.2   Euler angles (α, β, γ) for roll–yaw–roll geometry.

$$R_{x'}(\beta) = \begin{bmatrix} 1 & 0 & 0 \\ 0 & \cos\beta & -\sin\beta \\ 0 & \sin\beta & \cos\beta \end{bmatrix}_{\text{yaw}}$$

(3.3)

$$R_{z''}(\gamma) = \begin{bmatrix} \cos\gamma & -\sin\gamma & 0 \\ \sin\gamma & \cos\gamma & 0 \\ 0 & 0 & 1 \end{bmatrix}_{\text{roll}}$$

(3.4)

Note that 1 in the matrices above corresponds to the axis about which the rotation is performed.

The matrix product $R_{RYR}(\alpha, \beta, \gamma)$ in expanded form results in Equation 3.5. Consult the trigonometric formulas in Appendix 2.A.

$$R_{RYR}(\alpha, \beta, \gamma) =$$

$$\begin{bmatrix} \cos\alpha\cos\gamma - \sin\alpha\cos\beta\sin\gamma & -\cos\alpha\sin\gamma - \sin\alpha\cos\beta\cos\gamma & \sin\alpha\sin\beta \\ \sin\alpha\cos\gamma + \cos\alpha\cos\beta\sin\gamma & -\sin\alpha\sin\gamma + \cos\alpha\cos\beta\cos\gamma & -\cos\alpha\sin\beta \\ \sin\beta\sin\gamma & \sin\beta\cos\gamma & \cos\beta \end{bmatrix}$$

$$(3.5)$$

A special roll–yaw–roll geometry is obtained when the rotation matrix $R_{RYR}(\alpha, \beta, \gamma)$ becomes degenerate, that is. when yaw rotation is equal to

$$\beta = 0° \text{ or } \beta = 180°$$

(1) When $\beta = 0°$, the z and z'' axes become aligned and parallel to each other, both α and γ rotations are taking place about the same axis and then the roll–yaw–roll matrix in Equation 3.5 yields

$$R_{RYR}(\alpha, \beta, \gamma) = \begin{bmatrix} \cos(\alpha+\gamma) & -\sin(\alpha+\gamma) & 0 \\ \sin(\alpha+\gamma) & \cos(\alpha+\gamma) & 0 \\ 0 & 0 & 1 \end{bmatrix}_{\beta=0°} \qquad (3.6)$$

(2) When $\beta = 180°$, the z and z'' axes are aligned but antiparallel, thus the roll–yaw–roll matrix in Equation 3.5 yields

$$R_{RYR}(\alpha, \beta, \gamma) = \begin{bmatrix} \cos(\alpha-\gamma) & \sin(\alpha-\gamma) & 0 \\ \sin(\alpha-\gamma) & -\cos(\alpha-\gamma) & 0 \\ 0 & 0 & -1 \end{bmatrix}_{\beta=180°} \qquad (3.7)$$

We can restrict to the range $(0°, 180°)$ since beyond this range we obtain no new rotation matrices. The matrix specified for the range $(\alpha + 180°, -\beta, \gamma + 180°)$ is the same as the $R_{RYR}(\alpha, \beta, \gamma)$ matrix for positive β.

(2) Roll–pitch–yaw geometry

The roll–pitch–yaw geometry is the most comprehensive robot convention, as all three rotations in their reference positions are involved. Roll corresponds to the rotation about the z axis, pitch means the rotation about the new y axis designated by y' and yaw corresponds to the rotation about the new x' axis designated by x''. The roll–pitch–yaw geometry is illustrated with its corresponding Euler angles in Figure 3.3.

The matrix of the overall rotation $R_{RPY}(\phi, \theta, \psi)$ is clearly the product of three single axis rotation matrices corresponding to the individual axis rotations ϕ, θ and ψ:

$$R_{RPY}(\phi, \theta, \psi) = R_z(\phi) * R_{y'}(\theta) * R_{x''}(\psi) \tag{3.8}$$

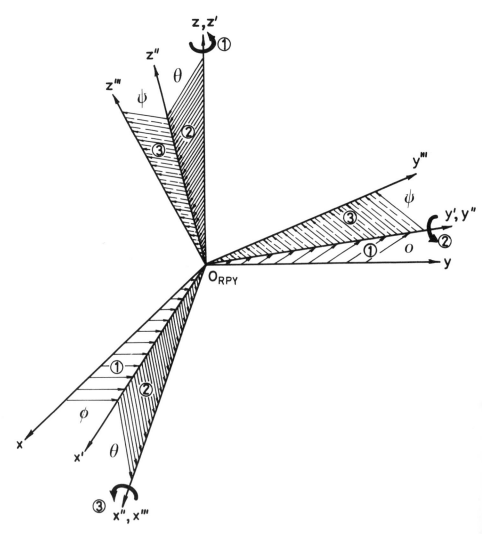

Figure 3.3 Euler angles (ϕ, θ, ψ) for roll–pitch–yaw geometry.

where

ϕ = rotation about the z axis – roll
θ = rotation about the y' axis – pitch
ψ = rotation about the x'' axis – yaw

The rotation matrices corresponding to each of the single axis rotations are as follows:

$$R_z(\phi) = \begin{bmatrix} \cos\phi & -\sin\phi & 0 \\ \sin\phi & \cos\phi & 0 \\ 0 & 0 & 1 \end{bmatrix}_{\text{roll}} \tag{3.9}$$

$$R_{y'}(\theta) = \begin{bmatrix} \cos\theta & 0 & \sin\theta \\ 0 & 1 & 0 \\ -\sin\theta & 0 & \cos\theta \end{bmatrix}_{\text{pitch}} \tag{3.10}$$

$$R_{x''}(\psi) = \begin{bmatrix} 1 & 0 & 0 \\ 0 & \cos\psi & -\sin\psi \\ 0 & \sin\psi & \cos\psi \end{bmatrix}_{\text{yaw}} \tag{3.11}$$

The matrix product $R_{\text{RPY}}(\phi, \theta, \psi)$ in Equation 3.8 in expanded form results in

$R_{\text{RPY}}(\phi, \theta, \psi)$

$$= \begin{bmatrix} \cos\phi\cos\theta & \cos\phi\sin\theta\sin\psi - \sin\phi\cos\psi & \cos\phi\sin\theta\cos\psi + \sin\phi\sin\psi \\ \sin\phi\cos\theta & \sin\phi\sin\theta\sin\psi + \cos\phi\cos\psi & \sin\phi\sin\theta\cos\psi - \cos\phi\sin\psi \\ -\sin\theta & \cos\theta\sin\psi & \cos\theta\cos\psi \end{bmatrix}$$

$$\tag{3.12}$$

A special roll–pitch–yaw geometry is obtained when the rotation matrix $R_{\text{RPY}}(\phi, \theta, \psi)$ becomes degenerate, that is when pitch rotation is equal to

$$\theta = +90° \quad \text{or} \quad \theta = -90°$$

(1) When $\theta = 90°$, the x'' axis aligns with the z axis and is antiparallel, both ϕ and ψ rotations are performed about the same axis and the roll–pitch–yaw matrix in Equation 3.12 yields

$$R_{RPY}(\phi, \theta, \psi) = \begin{bmatrix} 0 & \sin(\psi - \phi) & \cos(\psi - \phi) \\ 0 & \cos(\psi - \phi) & -\sin(\psi - \phi) \\ -1 & 0 & 0 \end{bmatrix}_{\theta = 90°} \tag{3.13}$$

(2) When $\theta = -90°$, the x'' and z axes are aligned and parallel, thus the roll–pitch–yaw matrix in Equation 3.12 yields

$$R_{RPY}(\phi, \theta, \psi) = \begin{bmatrix} 0 & -\sin(\psi + \phi) & -\cos(\psi + \phi) \\ 0 & \cos(\psi + \phi) & -\sin(\psi + \phi) \\ 1 & 0 & 0 \end{bmatrix}_{\theta = -90°} \tag{3.14}$$

We can restrict to the range of $(-90°, 90°)$ since beyond this range we obtain no new rotation matrices. The matrix specified for $(\phi - 90°, -\theta, \psi + 90°)$ is the same as the R_{RPY} matrix for positive θ.

(3) Roll–pitch–roll geometry

In the roll–pitch–roll convention, roll stands for rotation about the z axis, pitch about the new y axis designated by y' and roll again about the new z axis designated by z''. The roll–pitch–roll geometry is illustrated with its corresponding Euler angles in Figure 3.4.

    The overall rotation matrix $R_{RPR}(\delta, \lambda, \sigma)$ is the product of the three single axis rotation matrices corresponding to the individual rotation angles δ, λ and σ.

$$R_{RPR}(\delta, \lambda, \sigma) = R_z(\delta) * R_{y'}(\lambda) * R_{z''}(\sigma) \tag{3.15}$$

where

$\delta = $ rotation about the z axis – roll
$\lambda = $ rotation about the y' axis – pitch
$\sigma = $ rotation about the z'' axis – roll

The rotation matrices corresponding to the single axis rotations are as follows:

$$R_z(\delta) = \begin{bmatrix} \cos \delta & -\sin \delta & 0 \\ \sin \delta & \cos \delta & 0 \\ 0 & 0 & 1 \end{bmatrix}_{roll} \tag{3.16}$$

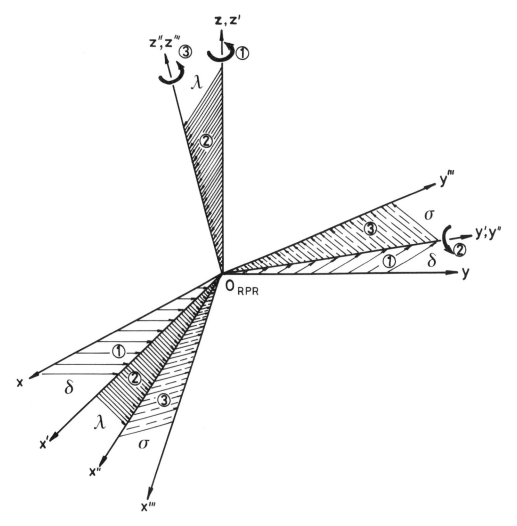

Figure 3.4   Euler angles (δ, λ, σ) for roll–pitch–roll robot geometry.

$$R_{y'}(\lambda) = \begin{bmatrix} \cos \lambda & 0 & \sin \lambda \\ 0 & 1 & 0 \\ -\sin \lambda & 0 & \cos \lambda \end{bmatrix}_{\text{pitch}} \tag{3.17}$$

$$R_{z''}(\sigma) = \begin{bmatrix} \cos \sigma & -\sin \sigma & 0 \\ \sin \sigma & \cos \sigma & 0 \\ 0 & 0 & 1 \end{bmatrix}_{\text{roll}} \tag{3.18}$$

The matrix product $R_{RPR}(\delta, \lambda, \sigma)$ in expanded form results in Equation 3.19 as

$$R_{RPR}(\delta, \lambda, \sigma) =$$

$$\begin{bmatrix} \cos\delta\cos\lambda\cos\sigma - \sin\delta\sin\sigma & -\cos\delta\cos\lambda\sin\sigma - \sin\delta\cos\sigma & \cos\delta\sin\lambda \\ \sin\delta\cos\lambda\cos\sigma + \cos\delta\sin\sigma & -\sin\delta\cos\lambda\sin\sigma + \cos\delta\cos\sigma & \sin\delta\sin\lambda \\ -\sin\lambda\cos\sigma & \sin\lambda\sin\sigma & \cos\lambda \end{bmatrix}$$

$$(3.19)$$

A special roll–pitch–roll geometry is obtained when the rotation matrix $R_{RPR}(\delta, \lambda, \sigma)$ becomes degenerate, that is when the pitch rotation is equal to

$$\lambda = 0° \text{ or } \lambda = 180°$$

(1) When $\lambda = 0°$, the δ and σ rotations are performed about the same axis in the same direction. In this case Equation 3.19 becomes

$$R_{RPR}(\delta, \lambda, \sigma) = \begin{bmatrix} \cos(\delta + \sigma) & -\sin(\delta + \sigma) & 0 \\ \sin(\delta + \sigma) & \cos(\delta + \sigma) & 0 \\ 0 & 0 & 1 \end{bmatrix}_{\lambda = 0°} \qquad (3.20)$$

(2) When $\lambda = 180°$, then the δ and σ rotations are performed about the same axis in the opposite direction. In this case Equation 3.19 becomes

$$R_{RPR}(\delta, \lambda, \sigma) = \begin{bmatrix} -\cos(\delta - \sigma) & -\sin(\delta - \sigma) & 0 \\ -\sin(\delta - \sigma) & \cos(\delta - \sigma) & 0 \\ 0 & 0 & -1 \end{bmatrix}_{\lambda = 180°} \qquad (3.21)$$

In conclusion, given the three corresponding angles for any configuration of rotations, one can easily derive the uniquely determined rotation matrices of Equations 3.5, 3.12 and 3.19. In these equations there are three sine and three cosine evaluations, fourteen multiplications, four additions/subtractions, all in all twenty-four computational calculations. It is worthwhile mentioning that the computational efficiency of a method is estimated in terms of the number of multiplications, divisions, additions/subtractions, calculation of trigonometric functions.

3.2   EULER ANGLES FOR ROLL–YAW–ROLL GEOMETRY

After having performed a series of homogeneous transformations on an object, the task undertaken in robot operation design is to determine the orientation of an object in terms of Euler angles. The solution of the roll–yaw–roll geometry is obtained by equating Equations 2.1 and 3.5, which then yields

$$
\begin{bmatrix}
e_{1x} & e_{2x} & e_{3x} \\
e_{1y} & e_{2y} & e_{3y} \\
e_{1z} & e_{2z} & e_{3z}
\end{bmatrix} =
$$

$$
\begin{bmatrix}
\cos\alpha\cos\gamma - \sin\alpha\cos\beta\sin\gamma & -\cos\alpha\sin\gamma - \sin\alpha\cos\beta\cos\gamma & \sin\alpha\sin\beta \\
\sin\alpha\cos\gamma + \cos\alpha\cos\beta\sin\gamma & -\sin\alpha\sin\gamma + \cos\alpha\cos\beta\cos\gamma & -\cos\alpha\sin\beta \\
\sin\beta\sin\gamma & \sin\beta\cos\gamma & \cos\beta
\end{bmatrix}
$$

$$(3.22)$$

Equation 3.22 provides nine nontrivial equations for the three unknown angles α, β, γ. One set of the solutions is straightforward. Let us equate, for instance, the corresponding terms in Equation 3.22, as follows:

$$e_{3z} = \cos\beta \qquad \rightarrow \beta = \arccos e_{3z} \tag{3.23}$$

$$e_{3y} = -\cos\alpha\sin\beta \rightarrow \alpha = \arccos\left(\frac{-e_{3y}}{\sin\beta}\right) \tag{3.24}$$

$$e_{2z} = \sin\beta\cos\gamma \qquad \rightarrow \gamma = \arccos\left(\frac{e_{2z}}{\sin\beta}\right) \tag{3.25}$$

This straightforward derivation of the Euler angles does not lead to a satisfactory solution, and may cause large errors for the following reasons.

(1) Equations 3.24 and 3.25 are undefined for $\beta = 0°$ and $\beta = 180°$, i.e. these are the special cases in Equations 3.6 and 3.7.
(2) When using the arccosine function in Equation 3.23 for obtaining an angle, not only the sign of the angle is undefined, but the accuracy in calculating the angle depends upon the value of that angle.
(3) In Equations 3.24 and 3.25 we again use the arccosine function which is divided by a sine function which may cause further deterioration of accuracy.

In order to avoid the errors in calculating the angles outlined above, we will use, throughout this book, the arctan function which is given by two arguments, say x and y. This function returns angles in the range $(-180°, 180°)$

by examining the sign of both arguments. In addition it also detects when either x or y is zero and then returns the correct values. Its accuracy will remain the same over its full defined range. In most computer systems there is a facility for calculating arctan functions which have the properties above.

In calculating the Euler angles α, β, γ for the roll–yaw–roll configuration by using this improved computing stability, we now choose the relationships in Equation 3.22 in the following way:

$$\cos \beta = e_{3z} \tag{3.26}$$

$$\sin^2 \beta = \frac{e_{3x}^2 + e_{3y}^2 + e_{1z}^2 + e_{2z}^2}{2} \tag{3.27}$$

and therefore

$$\beta = \arctan \frac{\sqrt{\tfrac{1}{2}(e_{3x}^2 + e_{3y}^2 + e_{1z}^2 + e_{2z}^2)}}{e_{3z}} \tag{3.28}$$

If the value of Equation 3.27 is nonzero, i.e. $\sin^2 \beta \neq 0$, then we can use

$$\sin \alpha \sin \beta = e_{3x} \tag{3.29}$$

$$\cos \alpha \sin \beta = -e_{3y} \tag{3.30}$$

and therefore

$$\alpha = \arctan \left(\frac{-e_{3x}}{e_{3y}} \right) \tag{3.31}$$

In order to get higher computing stability for α, a more extensive relationship is provided (for proof see Appendix 3.A):

$$\alpha = \arctan \frac{e_{2z}e_{1y} - e_{2y}e_{1z}}{e_{1x}e_{2z} - e_{1z}e_{2x}} \tag{3.32}$$

Similarly

$$\sin \beta \sin \gamma = e_{1z} \tag{3.33}$$

$$\sin \beta \cos \gamma = e_{2z} \tag{3.34}$$

and therefore

$$\gamma = \arctan \frac{e_{1z}}{e_{2z}} \tag{3.35}$$

In the case of applying higher computing stability, γ can be related to a more extensive expression (see Appendix 3.A).

$$\gamma = \arctan \frac{e_{3x}e_{2y} - e_{2x}e_{3y}}{e_{1x}e_{3y} - e_{3x}e_{1y}} \tag{3.36}$$

A special geometric configuration for $\beta = 0°$ and $\beta = 180°$ is as follows:

(1) When $\beta = 0°$, i.e. Equation 3.26 gives the value 1, there is another way to derive α and γ, since they are no longer uniquely determinable. We can equate the corresponding members of Equation 3.22 as follows:

$$(1 + \cos \beta) \, \sin(\alpha + \gamma) = e_{1y} - e_{2x} \tag{3.37a}$$

$$(1 + \cos \beta) \cos(\alpha + \gamma) = e_{1x} + e_{2y} \tag{3.37b}$$

and then by dividing them, we get

$$\alpha + \gamma = \arctan \frac{e_{1y} - e_{2x}}{e_{1x} + e_{2y}} \tag{3.38}$$

(2) When $\beta = 180°$, then Equation 3.26 gives the value -1, and similarly, as above, the following relationships may be used

$$(1 - \cos \beta) \sin(\alpha - \gamma) = -e_{1y} - e_{2x} \tag{3.39a}$$

$$(1 - \cos \beta) \cos(\alpha - \gamma) = -e_{1x} - e_{2y} \tag{3.39b}$$

and then by dividing them, we get:

$$\alpha - \gamma = \arctan \frac{-e_{1y} - e_{2x}}{-e_{1x} + e_{2y}} \tag{3.40}$$

From Equations 3.38 and 3.40, the values of α and γ are readily available.

3.3   *ILLUSTRATIVE EXAMPLE 4* GRIPPER POSITIONING BY EULER ANGLES FOR ROLL–YAW–ROLL GEOMETRY

Specification

The S plane in Figure 3.5 is defined by the points A(4, 0, 0), B(0, 6, 0), C(0, 0, 3). A robot manipulator is resting in its starting position $P_1(0, 0, 10)$ and pointing

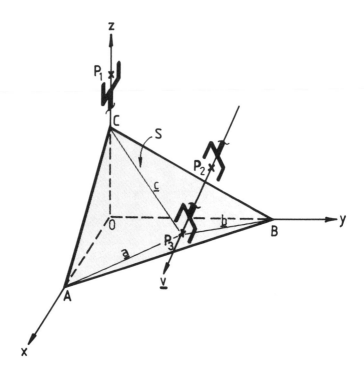

Figure 3.5   Gripper positioning in the workspace for roll–yaw–roll configuration.

in the z direction. The plane of the fingers of the gripper is parallel with the plane (x, z). The task is to guide the robot manipulator from P_1 through P_2 to P_3, where P_2 is an intermediate so called approach point and has coordinates $(4, 4, 3)$. The point P_3 lies in the S plane. The 'approach vector' $v = \overline{P_2 P_3}$ is perpendicular to the S plane. The plane of the fingers in the positions P_2 and P_3 is parallel with the z axis.

Design objectives

   (i) Calculate the cartesian coordinates of the gripper's center point in the position P_3.

   (ii) Derive the Euler angles α, β and γ of the gripper for roll–yaw–roll robot geometry in the location P_3.

   (iii) Develop the homogeneous transformation matrices which describe the motion of the gripper from the reference location P_1 through the interim position P_2 to location P_3, i.e. determine $H_{0,1}, H_{0,2}, H_{1,2}$ and $H_{2,3}$.

Solutions

(i) *Calculation of the cartesian coordinates of location P_3*
We need first to provide the equation of the S plane, which is specified when one point on it is known, together with any vector perpendicular to it. Such a vector, when normalized, is a unit normal to the plane and is unique except for its sign. The ambiguity as to the sign of the normal is, of course, because a plane has no preferred side. The cartesian equation of the S plane (see Appendix 3B) is given in a standard form [3.2] as

$$k_1 x + k_2 y + k_3 z = 1 \tag{3.41}$$

By substituting the coordinates of points A, B and C we get the equation

$$\tfrac{1}{4}x + \tfrac{1}{6}y + \tfrac{1}{3}z = 1 \tag{3.42}$$

or

$$3x + 2y + 4z = 12 \tag{3.43}$$

where the coefficients of x, y and z are the corresponding components of the normal vector to S, i.e. the components of vector v respectively. Since vector v is perpendicular to S, it is also perpendicular to every vector being in the S plane and going through P_3. Thus v is also perpendicular to the three vectors in the S plane pointing from A, B and C to P_3 which are designated by a, b and c respectively. The condition of perpendicularity is that the scalar products of the vectors a, b and c with vector v must be zero.

$$a = \begin{bmatrix} x_3 - 4 \\ y_3 \\ z_3 \end{bmatrix} \quad b = \begin{bmatrix} x_3 \\ y_3 - 6 \\ z_3 \end{bmatrix} \quad c = \begin{bmatrix} x_3 \\ y_3 \\ z_3 - 3 \end{bmatrix}$$

and the vector

$$v = \begin{bmatrix} 4 - x_3 \\ 4 - y_3 \\ 3 - z_3 \end{bmatrix} \tag{3.44}$$

The conditions of perpendicularity in mathematical terms are as follows:

$$a^{\mathrm{T}} v = 0$$

$$(x_3 - 4)(4 - x_3) + y_3(4 - y_3) + z_3(3 - z_3) = 0 \tag{3.45}$$

$$b^T v = 0$$

$$x_3(4 - x_3) + (y_3 - 6)(4 - y_3) + z_3(3 - z_3) = 0 \tag{3.46}$$

$$c^T v = 0$$

$$x_3(4 - x_3) + y_3(4 - y_3) + (z_3 - 3)(3 - z_3) = 0 \tag{3.47}$$

By substituting the coordinates of P_3 in Equation 3.43, the cartesian equation of the S plane becomes

$$3x_3 + 2y_3 + 4z_3 = 12 \tag{3.48}$$

By taking the difference of Equations 3.45 and 3.46 as well as that of Equations 3.45 and 3.47 and expressing y_3 and z_3 in terms of x_3 and substituting them into Equation 3.48, we obtain

$$3x_3 + \left(\frac{8}{3} + \frac{4}{3}x_3\right) + \left(-\frac{28}{3} + \frac{16}{3}x_3\right) = 12 \tag{3.49}$$

and from this the cartesian coordinates of the tool center point in the position of P_3 are

$$x_3 = \frac{56}{29} = 1.93 \quad y_3 = \frac{76}{29} = 2.62 \quad z_3 = \frac{7}{29} = 0.24 \tag{3.50}$$

(ii) *Calculation of the Euler angles α, β and γ*
In the reference position of the gripper the vector pointing from one fingertip to the other is in the x direction (see Figure 3.5). The geometry of Euler angles β and α is shown in Figures 3.6(a) and 3.6(b) respectively.

The angle α is between the reference finger and the actual finger vectors of the gripper at P_3

$$\alpha = \arctan \frac{y_2 - y_3}{x_2 - x_3} + 180° \tag{3.51a}$$

and substituting the corresponding values of the coordinates into Equation 3.51a yields

$$\alpha = \arctan\left(\frac{4 - \frac{76}{29}}{4 - \frac{56}{29}}\right) + 180° = 213.7° \tag{3.51b}$$

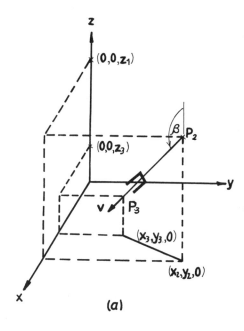

(a)

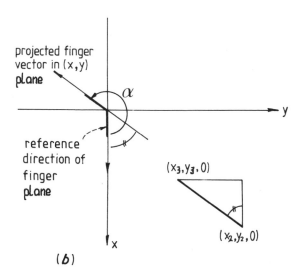

(b)

Figure 3.6   Geometry of Euler angles. (a) β-angle. (b) α-angle.

The angle β is the angle between the directions of the z axis and $\mathbf{v}$ vector

$$\beta = 180° - \arctan \frac{\sqrt{(x_2 - x_3)^2 + (y_2 - y_3)^2}}{z_2 - z_3} \qquad (3.52a)$$

and substituting the corresponding values of the coordinates into Equation 3.52a

$$\beta = 180° - \arctan \frac{\sqrt{(4-\frac{56}{29})^2 + (4-\frac{76}{29})^2}}{3-\frac{7}{29}} = 138° \tag{3.52b}$$

Finally

$$\gamma = 0°$$

since the rotation took place in the plane of the fingers which remained parallel with the z axis.

(iii) *Calculation of the transformation matrices between the locations P_1 and P_2 and then P_2 and P_3*

The state of the gripper in its reference position P_1 is given by

$$\mathbf{H}_{0,1} = \begin{bmatrix} 1 & 0 & 0 & 0 \\ 0 & 1 & 0 & 0 \\ 0 & 0 & 1 & 10 \\ 0 & 0 & 0 & 1 \end{bmatrix} \tag{3.53}$$

The state of the gripper in its interim position P_2 is derived from Equation 3.5 and given by

$$\mathbf{H}_{0,2} = \begin{bmatrix} \cos\alpha & -\sin\alpha\cos\beta & \sin\alpha\sin\beta & x_1 \\ \sin\alpha & \cos\alpha\cos\beta & -\cos\alpha\sin\beta & y_1 \\ 0 & \sin\beta & \cos\beta & z_1 \\ 0 & 0 & 0 & 1 \end{bmatrix} \tag{3.54}$$

where $\gamma = 0°$ has been taken into consideration

$$\mathbf{H}_{0,2} = \begin{bmatrix} -0.83 & -0.41 & -0.37 & 4 \\ -0.55 & 0.61 & 0.56 & 4 \\ 0 & 0.67 & -0.74 & 3 \\ 0 & 0 & 0 & 1 \end{bmatrix} \tag{3.55}$$

For the location P_3 the rotation submatrix remained unchanged, only the translation column in Equation 3.56 has new values obtained in Equation 3.50.

$$H_{0,3} = \begin{bmatrix} -0.83 & -0.41 & -0.37 & 1.93 \\ -0.55 & 0.61 & 0.56 & 2.62 \\ 0 & 0.67 & -0.74 & 0.24 \\ 0 & 0 & 0 & 1 \end{bmatrix} \tag{3.56}$$

The transformation matrix from P_1 to P_2 is obtainable from Equations (3.53) and (3.55). (For matrix inversion see Appendix 4.C.)

$$H_{1,2} = H_{0,1}^{-1} * H_{0,2}$$

$$H_{1,2} = \begin{bmatrix} 1 & 0 & 0 & 0 \\ 0 & 1 & 0 & 0 \\ 0 & 0 & 1 & -10 \\ 0 & 0 & 0 & 1 \end{bmatrix} * \begin{bmatrix} -0.83 & -0.41 & -0.37 & 4 \\ -0.55 & 0.61 & 0.56 & 4 \\ 0 & 0.67 & -0.74 & 3 \\ 0 & 0 & 0 & 1 \end{bmatrix}$$

$$= \begin{bmatrix} -0.83 & -0.41 & -0.37 & 4 \\ -0.55 & 0.61 & 0.56 & 4 \\ 0 & 0.67 & -0.74 & -7 \\ 0 & 0 & 0 & 1 \end{bmatrix} \tag{3.57}$$

Since the orientation of the gripper does not change between P_2 and P_3 the transformation matrix describing the motion from P_2 to P_3 will be

$$H_{2,3} = H_{0,2}^{-1} * H_{0,3}$$

$$H_{2,3} = \begin{bmatrix} 1 & 0 & 0 & 2.07 \\ 0 & 1 & 0 & 1.38 \\ 0 & 0 & 1 & 2.76 \\ 0 & 0 & 0 & 1 \end{bmatrix} \tag{3.58}$$

That is the displacement between P_2 and P_3 is

$$\Delta x = 2.07 \quad \Delta y = 1.38 \quad \Delta z = 2.76$$

3.4   EULER ANGLES FOR ROLL–PITCH–YAW GEOMETRY

Euler angles for roll–pitch–yaw geometry are deduced from Equations 2.1 and 3.12 and we then have

$$
\begin{bmatrix}
e_{1x} & e_{2x} & e_{3x} \\
e_{1y} & e_{2y} & e_{3y} \\
e_{1z} & e_{2z} & e_{3z}
\end{bmatrix}
$$

$$
= \begin{bmatrix}
\cos\phi\cos\theta & \cos\phi\sin\theta\sin\psi - \sin\phi\cos\psi & \cos\phi\sin\theta\cos\psi + \sin\phi\sin\psi \\
\sin\phi\cos\theta & \sin\phi\sin\theta\sin\psi + \cos\phi\cos\psi & \sin\phi\sin\theta\cos\psi - \cos\phi\sin\psi \\
-\sin\theta & \cos\theta\sin\psi & \cos\theta\cos\psi
\end{bmatrix}
$$

$$(3.59)$$

As explained in Section 3.2, a direct solution for the angles ϕ, θ and ψ by using e_{1z}, e_{1y} and e_{2z} in Equation 3.59 is of no use and so we have to express the tangents of the unknowns to achieve numerical and computational stability:

$$\sin\theta = -e_{1z} \tag{3.60}$$

$$\cos^2\theta = \frac{e_{1x}^2 + e_{1y}^2 + e_{2z}^2 + e_{3z}^2}{2} \tag{3.61}$$

and

$$\theta = \arctan \frac{-e_{1z}}{\sqrt{\frac{1}{2}(e_{1z}^2 + e_{1y}^2 + e_{2z}^2 + e_{3z}^2)}} \tag{3.62}$$

If the value of $\cos^2\theta$ in Equation 3.61 is nonzero, the angle ϕ is obtained (see Appendix 3.C) as

$$\phi = \arctan \frac{e_{2z}e_{3x} - e_{3z}e_{2x}}{e_{3z}e_{2y} - e_{2z}e_{3y}} \tag{3.63}$$

Similarly (see Appendix 3.C)

$$\psi = \arctan \frac{e_{1y}e_{3x} - e_{1x}e_{3y}}{e_{1x}e_{2y} - e_{1y}e_{2x}} \tag{3.64}$$

A special geometry configuration for $\theta = 90°$ and $\theta = -90°$ is as follows:

(1) If $\theta = 90°$ then $\sin \theta$ in Equation 3.60 yields -1, then we have

$$e_{2x} - e_{3y} = (1 + \sin \theta)\cos \phi \sin \psi - (1 + \sin \theta)\sin \phi \cos \psi$$
$$= (1 + \sin \theta)\sin(\psi - \phi)$$
$$e_{2y} + e_{3x} = (1 + \sin \theta)\sin \phi \sin \psi + (1 + \sin \theta)\cos \phi \cos \psi$$
$$= (1 + \sin \theta)\cos(\psi - \phi)$$

and we get

$$\psi - \theta = \arctan \frac{e_{2x} - e_{3y}}{e_{2y} + e_{3x}} \tag{3.65}$$

(2) If $\theta = -90°$, then $\sin \theta$ in Equation 3.60 yields $+1$, and we similarly get as above

$$\psi + \phi = \arctan \frac{-e_{2x} - e_{3y}}{e_{2y} - e_{3x}} \tag{3.66}$$

3.5   *ILLUSTRATIVE EXAMPLE 5* GRIPPER POSITIONING BY EULER ANGLES FOR ROLL–PITCH–YAW GEOMETRY

Specification

The specification is the same as it was in Illustrative example 4. In fact this exercise is an extension for the calculation of Euler angles of the gripper.

Design objectives

Derive the Euler angles of the gripper in the position P_3 for roll–pitch–yaw geometry configuration, i.e. what are the values of angles ϕ, θ and ψ (see Figure 3.5).

Solution

The state of the gripper at P_3 is given by Equation 3.56, the upper left 3×3 submatrix of which provides the orientation matrix R_3 of the gripper.

$$R_3 = \begin{bmatrix} -0.83 & -0.41 & -0.37 \\ -0.55 & 0.61 & 0.56 \\ 0 & 0.67 & -0.74 \end{bmatrix} \tag{3.67}$$

By considering the element in R_3, we see that for $\theta = 0°$, $\sin \theta = 0$ and $\cos \theta = 1$, which can be checked against the sum $(e_{1x}^2 + e_{1z}^2)$,

$$\phi = \arctan \frac{e_{1y}}{e_{1x}} = \arctan \frac{-0.55}{-0.83} = 33.5° \tag{3.68}$$

$$\psi = \arctan \frac{e_{2z}}{e_{3z}} = \arctan \frac{0.67}{-0.74} = -42.2° \tag{3.69}$$

The correct solution for ψ is given by (see Equation 3.67)

$$\psi = 180° - 42.2° = 138.8°$$

since $e_{2z} = \sin \psi > 0$ and $e_{3z} = \cos \psi < 0$. But these are possible only if a further condition for ψ is satisfied, that is

$$90° < \psi < 180°$$

3.6   EULER ANGLES FOR ROLL–PITCH–ROLL GEOMETRY

Euler angles for roll–pitch–roll geometry are derived by equating Equations 2.1 and 3.19. Then we have

$$\begin{bmatrix} e_{1x} & e_{2x} & e_{3x} \\ e_{1y} & e_{2y} & e_{3y} \\ e_{1z} & e_{2z} & e_{3z} \end{bmatrix}$$

$$= \begin{bmatrix} \cos\delta\cos\lambda\cos\sigma - \sin\delta\sin\sigma & -\cos\delta\cos\lambda\sin\sigma - \sin\delta\cos\sigma & \cos\delta\sin\lambda \\ \sin\delta\cos\lambda\cos\sigma + \cos\delta\sin\sigma & -\sin\delta\cos\lambda\sin\sigma + \cos\delta\cos\sigma & \sin\delta\sin\lambda \\ -\sin\lambda\cos\sigma & \sin\lambda\sin\sigma & \cos\lambda \end{bmatrix} \tag{3.70}$$

$$\cos\lambda = e_{3z} \tag{3.71}$$

$$\sin^2\lambda = \frac{e_{3x}^2 + e_{3y}^2 + e_{1z}^2 + e_{2z}^2}{2} \tag{3.72}$$

and then by dividing the square root of Equation 3.72 by Equation 3.71 we obtain

$$\lambda = \arctan \frac{\sqrt{\frac{1}{2}(e_{3x}^2 + e_{3y}^2 + e_{1z}^2 + e_{2z}^2)}}{e_{3z}} \tag{3.73}$$

If the value of Equation 3.72 is nonzero, i.e. $\sin^2 \lambda \neq 0$, then we can use

$$\sin \delta \sin \lambda = e_{3y} \tag{3.74}$$

$$\cos \delta \sin \lambda = e_{3x} \tag{3.75}$$

and therefore

$$\delta = \arctan \frac{e_{3y}}{e_{3x}} \tag{3.76}$$

or in order to get higher numerical stability we use

$$\cos \delta \sin \lambda = e_{2z}e_{1y} + e_{1z}e_{2y} \tag{3.77}$$

$$\sin \delta \sin \lambda = -(e_{2z}e_{1x} + e_{1z}e_{2x}) \tag{3.78}$$

and therefore

$$\delta = \arctan \left(-\frac{e_{2z}e_{1x} + e_{1z}e_{2x}}{e_{2z}e_{1y} + e_{1z}e_{2y}} \right) \tag{3.79}$$

Similarly

$$-\sin \lambda \cos \sigma = e_{1z} \tag{3.80}$$

$$\sin \lambda \sin \sigma = e_{2z} \tag{3.81}$$

and therefore

$$\sigma = \arctan -\frac{e_{2z}}{e_{1z}} \tag{3.82}$$

or again to gain higher stability:

$$\sin \delta \cos \sigma = e_{3x}e_{2y} - e_{2x}e_{3y} \tag{3.83}$$

$$\sin \delta \sin \sigma = e_{3x}e_{1y} - e_{1x}e_{3y} \tag{3.84}$$

and

$$\sigma = \arctan\left(\frac{e_{3x}e_{1y} - e_{1x}e_{3y}}{e_{3x}e_{2y} - e_{2x}e_{3y}}\right) \tag{3.85}$$

A special geometric configuration for $\lambda = 0°$ and $\lambda = 180°$ is as follows:

(1) When $\lambda = 0°$, i.e. Equation 3.71 gives the value 1, we have to use another method since δ and σ are no longer uniquely determined. We can equate as follows

$$(1 + \cos \lambda)\sin(\delta + \sigma) = e_{1y} - e_{2x} \tag{3.86a}$$

$$(1 + \cos \lambda)\cos(\delta + \sigma) = e_{1x} + e_{2y} \tag{3.86b}$$

and then by dividing, we get

$$\delta + \sigma = \arctan\left(\frac{e_{1y} - e_{2x}}{e_{1x} + e_{2y}}\right) \tag{3.87}$$

(2) When $\lambda = 180°$ then in a way similar to that of Equations 3.86a and 3.86b the corresponding equations become

$$(1 - \cos \lambda)\sin(\delta - \sigma) = -e_{1y} - e_{2x} \tag{3.88a}$$

$$(1 - \cos \lambda)\cos(\delta - \sigma) = -e_{1x} + e_{2y} \tag{3.88b}$$

and by dividing, we get

$$\delta - \sigma = \arctan\left(\frac{-e_{1y} - e_{2x}}{-e_{1x} + e_{2y}}\right) \tag{3.89}$$

3.7   CYLINDRICAL ROBOT COORDINATES

In Sections 2.1 and 3.3 we expressed the spatial positions of objects in terms of cartesian coordinates, that is by their x, y and z components. We might, however, wish to specify the positions of objects in noncartesian coordinates, namely either in cylindrical or polar (spherical) coordinate frames [3.3].

In this section we will deal with the position of an object expressed in cylindrical coordinates corresponding to a translation h along the z_B axis, a rotation ρ about the z_B axis, and a translation r along the x_B axis (see Figure 3.7).

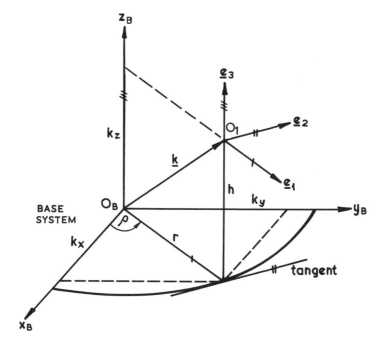

Figure 3.7 Cylindrical coordinates (h, ρ, r).

The homogeneous transformation expressing the effect of two translations and one rotation is

$$H_{\text{CYL}}(h, \rho, r) = H(h) * H(\rho) * H(r) \tag{3.90}$$

where

$$H(h) = \begin{bmatrix} 1 & 0 & 0 & 0 \\ 0 & 1 & 0 & 0 \\ 0 & 0 & 1 & h \\ 0 & 0 & 0 & 1 \end{bmatrix} \tag{3.91}$$

$$H(\rho) = \begin{bmatrix} \cos\rho & -\sin\rho & 0 & 0 \\ \sin\rho & \cos\rho & 0 & 0 \\ 0 & 0 & 1 & 0 \\ 0 & 0 & 0 & 1 \end{bmatrix} \tag{3.92}$$

and

$$H(r) = \begin{bmatrix} 1 & 0 & 0 & r \\ 0 & 1 & 0 & 0 \\ 0 & 0 & 1 & 0 \\ 0 & 0 & 0 & 1 \end{bmatrix}$$

(3.93)

Thus the homogeneous transformation in Equation 3.90 in matrix form becomes

$$H_{\text{CYL}} = \begin{bmatrix} e_{1x} & e_{2x} & e_{3x} & k_x \\ e_{1y} & e_{2y} & e_{3y} & k_y \\ e_{1z} & e_{2z} & e_{3z} & k_z \\ 0 & 0 & 0 & 1 \end{bmatrix} = \begin{bmatrix} \cos\rho & -\sin\rho & 0 & r\cos\rho \\ \sin\rho & \cos\rho & 0 & r\sin\rho \\ 0 & 0 & 1 & h \\ 0 & 0 & 0 & 1 \end{bmatrix}$$

(3.94)

Looking at the column vectors of H_{CYL} (h, ρ, r) we can see that e_1 is a normal vector to the cylinder of radius r, e_2 is a tangent vector to it and e_3 is a generatrix of the cylinder. All three intersect through a point defined by the translation coordinates (k_x, k_y, k_z). There is a wide range of robots whose basic motions are operated in cylindrical coordinates.

3.8 *ILLUSTRATIVE EXAMPLE 6*
CALCULATION OF CYLINDRICAL COORDINATES

Specification

A coordinate system of origin O_B in Figure 3.7 is given by the position vector k and the unitary axis vectors e_1, e_2, e_3 on the corresponding axes.

Design objectives

By knowing the homogeneous transformation matrix set up by the unitary axis vectors and by a position vector, let us calculate the cylindrical coordinates ρ, r and h resulting in a homogeneous transformation matrix by choosing a numerically stable and uniformly correct method.

Solutions

From Equation 3.94

$$k_x^2 + k_y^2 = (r \cos \rho)^2 + (r \sin \rho)^2 = r^2$$

where

$$r = \sqrt{k_x^2 + k_y^2}$$

and

$$h = k_z$$

if

$$e_{1x} = \cos \rho \neq 0$$

then

$$e_{1x}e_{2y} + e_{1y}e_{2x} = \cos^2 \rho - \sin^2 \rho = \cos 2\rho$$

$$e_{1x}e_{1y} - e_{2x}e_{2y} = 2 \sin \rho \cos \rho = \sin 2\rho$$

$$\rho = \tfrac{1}{2} \arctan \frac{e_{1x}e_{1y} - e_{2x}e_{2y}}{e_{1x}e_{2y} + e_{1y}e_{2x}}$$

if

$$e_{1x} = \cos \rho = 0$$

then

$$\rho = 90°$$

3.9   POLAR ROBOT COORDINATES

The other noncartesian coordinate system we will deal with is the polar (spherical) one. We would rather say polar than spherical, because for a spherical robot the three departed joints would be compressed into a single three degrees of freedom joint. An example of advantageous use of polar coordinates is the Unimation 2000 robot (see Section 1.5). Major axes of this robot correspond to polar coordinate axes. A position vector in a polar system is specified by a translation r in the direction of the z_B axis followed by a rotation ϑ about the y_B axis, and finally another rotation φ about the z_B axis (see Figure 3.8).

The homogeneous transformation expressing the effect of two rotations and one translation is presented as shown in Equations 3.95–3.98.

$$H_{POL}(\varphi, \vartheta, r) = H(\varphi) * H(\vartheta) * H(r) \tag{3.95}$$

where

$$H(\varphi) = \begin{bmatrix} \cos \varphi & -\sin \varphi & 0 & 0 \\ \sin \varphi & \cos \varphi & 0 & 0 \\ 0 & 0 & 1 & 0 \\ 0 & 0 & 0 & 1 \end{bmatrix} \tag{3.96}$$

$$H(\vartheta) = \begin{bmatrix} \cos \vartheta & 0 & \sin \vartheta & 0 \\ 0 & 1 & 0 & 0 \\ -\sin \vartheta & 0 & \cos \vartheta & 0 \\ 0 & 0 & 0 & 1 \end{bmatrix} \tag{3.97}$$

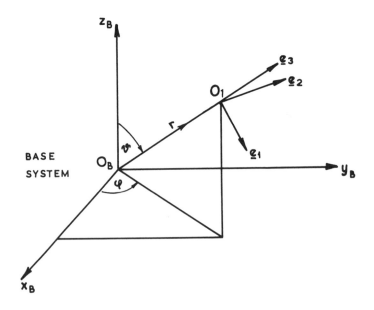

Figure 3.8   Polar coordinates (φ, ϑ, r).

$$H(r) = \begin{bmatrix} 1 & 0 & 0 & 0 \\ 0 & 1 & 0 & 0 \\ 0 & 0 & 1 & r \\ 0 & 0 & 0 & 1 \end{bmatrix} \tag{3.98}$$

Thus the resultant homogeneous transformation is

$$H_{\text{POL}}(\varphi, \vartheta, r) = \begin{bmatrix} e_{1x} & e_{2x} & e_{3x} & k_x \\ e_{1y} & e_{2y} & e_{3y} & k_y \\ e_{1z} & e_{2z} & e_{3z} & k_z \\ 0 & 0 & 0 & 1 \end{bmatrix}$$

$$= \begin{bmatrix} \cos\varphi\cos\vartheta & -\sin\varphi & \cos\varphi\sin\vartheta & r\cos\varphi\sin\vartheta \\ \sin\varphi\cos\vartheta & \cos\varphi & \sin\varphi\sin\vartheta & r\sin\varphi\sin\vartheta \\ -\sin\vartheta & 0 & \cos\vartheta & r\cos\vartheta \\ 0 & 0 & 0 & 1 \end{bmatrix}$$

$$\tag{3.99}$$

Let us now calculate the polar coordinates φ, ϑ, r from the given homogeneous transformation matrix $H_{\text{POL}}(\varphi, \vartheta, r)$. In order to get an explicit solution we premultiply Equation 3.95 by the inverse of one of the unknown transforms in Equations 3.96–3.98 [3.3], i.e. let us start with the inverse matrix $H(\varphi)^{-1}$.

$$H(\varphi)^{-1} * H_{\text{POL}}(\varphi, \vartheta, r) = H(\vartheta) * H(r) \tag{3.100}$$

After working out Equation 3.100 and by taking only the last columns on both sides, we obtain

$$\begin{bmatrix} k_x \cos\varphi + k_y \sin\varphi \\ -k_x \sin\varphi + k_y \cos\varphi \\ k_z \\ 1 \end{bmatrix} = \begin{bmatrix} r\sin\vartheta \\ 0 \\ r\cos\vartheta \\ 1 \end{bmatrix} \tag{3.101}$$

We get φ_1 and φ_2 from the second row in Equation 3.101

$$-k_x \sin\varphi + k_y \cos\varphi = 0 \tag{3.102}$$

$$\varphi_1 = \arctan \frac{k_y}{k_x} \tag{3.103}$$

$$\varphi_2 = \varphi_1 + 180° \tag{3.104}$$

Since, by definition, $r > 0$ we get from the first and third row of Equation 3.101

$$k_x \cos \varphi + k_y \sin \varphi = r \sin \vartheta \tag{3.105}$$

$$k_z = r \cos \vartheta \tag{3.106}$$

and

$$\vartheta = \arctan \frac{k_x \cos \varphi + k_y \sin \varphi}{k_z} \tag{3.107}$$

In order to derive the value of r, we premultiply Equation 3.100 by $H(\vartheta)^{-1}$ and obtain

$$H(\vartheta)^{-1} * H(\varphi)^{-1} * H_{POL}(\varphi, \vartheta, r) = H(r) \tag{3.108}$$

Since in Equation 3.98 r appears only in the last column of $H(r)$, we shall compare it only with the last column of Equation 3.108.

$$\begin{bmatrix} (k_x \cos \varphi + k_y \sin \varphi)\cos \vartheta - k_z \sin \vartheta \\ -k_x \sin \varphi + k_y \cos \varphi \\ (k_x \cos \varphi + k_y \sin \varphi)\sin \vartheta + k_z \cos \vartheta \\ 1 \end{bmatrix} = \begin{bmatrix} 0 \\ 0 \\ r \\ 1 \end{bmatrix} \tag{3.109}$$

that is

$$r = (k_x \cos \varphi + k_y \sin \varphi)\sin \vartheta + k_z \cos \vartheta \tag{3.110}$$

3.10   ILLUSTRATIVE EXAMPLE 7
CALCULATION OF CARTESIAN, CYLINDRICAL AND POLAR COORDINATES

Specification

As in Illustrative example 4.

Design objectives

Calculate the cartesian, cylindrical and polar coordinates of point P_3 in Figure 3.5.

Solutions

The coordinates of point P_3 in the three coordinate systems are given below. Compare the solution of the cartesian coordinates with that given in Equation 3.50.

Cartesian coordinates

$x_3 = 1.93$ (units)
$y_3 = 2.62$ (units)
$z_3 = 0.24$ (units)

Cylindrical coordinates

$$r_3 = \sqrt{x_3^2 + y_3^2} = 3.25 \text{ (units)}$$

$$\rho_3 = \arctan \frac{x_3}{y_3} = 36.38°$$

$$h_3 = 0.24 \text{ (units)}$$

Polar coordinates

$$r_3 = x_3^2 + y_3^2 + z_3^2 = 3.26 \text{ (units)}$$

$$\varphi_3 = \arctan \frac{x_3}{y_3} = 36.8°$$

$$\vartheta_3 = \arctan \frac{\sqrt{x_3 + y_3}}{z_3} = 85.78°$$

3.11   APPENDICES

Appendix 3.A

The derivation of Equations 3.32 and 3.36 is presented as follows:

Proof of Equation 3.32

$$e_{2z}e_{1y} = \sin \beta \cos \gamma (\sin \alpha \cos \gamma + \cos \alpha \cos \beta \sin \gamma)$$
$$= \sin \alpha \sin \beta \cos^2 \gamma + \cos \alpha \sin \beta \cos \beta \sin \gamma \cos \gamma \qquad (3.A.1)$$

$$e_{2y}e_{1z} = \sin \beta \sin \gamma(-\sin \alpha \sin \gamma + \cos \alpha \cos \beta \cos \gamma)$$
$$= -\sin \alpha \sin \beta \sin^2 \gamma + \cos \alpha \sin \beta \cos \beta \sin \gamma \cos \gamma \qquad (3.A.2)$$

Subtracting Equation 3.A.2 from Equation 3.A.1 yields

$$e_{2z}e_{1y} - e_{2y}e_{1z} = \sin \alpha \sin \beta(\cos^2 \gamma + \sin^2 \gamma)$$
$$= \sin \alpha \sin \beta \qquad (3.A.3)$$

☐ ☐ ☐

$$e_{1x}e_{2z} = \sin \beta \cos \gamma(\cos \alpha \cos \gamma - \sin \alpha \cos \beta \sin \gamma)$$
$$= \cos \alpha \sin \beta \cos^2 \gamma - \sin \alpha \sin \beta \cos \beta \sin \gamma \cos \gamma \qquad (3.A.4)$$

$$e_{1z}e_{2x} = \sin \beta \sin \gamma(-\cos \alpha \sin \gamma - \sin \alpha \cos \beta \cos \gamma)$$
$$= -\cos \alpha \sin \beta \sin^2 \gamma - \sin \alpha \sin \beta \cos \beta \sin \gamma \cos \gamma \qquad (3.A.5)$$

Subtracting Equation 3.A.5 from Equation 3.A.4 yields

$$e_{1x}e_{2z} - e_{1z}e_{2x} = \cos \alpha \sin \beta(\cos^2 \gamma + \sin^2 \gamma)$$
$$= \cos \alpha \sin \beta \qquad (3.A.6)$$

☐ ☐ ☐

Finally dividing Equation (3.A.3) by Equation 3.A.6 we get

$$\frac{e_{2z}e_{1y} - e_{2y}e_{1z}}{e_{1x}e_{2z} - e_{1z}e_{2x}} = \frac{\sin \alpha \sin \beta}{\cos \alpha \sin \beta} = \tan \alpha \qquad (3.A.7)$$

Provided that $\sin \beta$ is not equal to zero, α becomes

$$\alpha = \arctan \frac{e_{2z}e_{1y} - e_{2y}e_{1z}}{e_{1x}e_{2z} - e_{1z}e_{2x}} \qquad (3.32)$$

Proof of Equation 3.36

$$e_{3x}e_{2y} = \sin \alpha \sin \beta(-\sin \alpha \sin \gamma + \cos \alpha \cos \beta \cos \gamma)$$
$$= -\sin \beta \cos \gamma \sin^2 \alpha + \sin \alpha \sin \beta \cos \alpha \cos \beta \cos \gamma \qquad (3.A.8)$$

$$e_{2x}e_{3y} = \cos \alpha \sin \beta(-\cos \alpha \sin \gamma - \sin \alpha \cos \beta \cos \gamma)$$
$$= \sin \beta \sin \gamma \sin^2 \alpha + \sin \alpha \sin \beta \cos \alpha \cos \beta \cos \gamma \qquad (3.A.9)$$

Subtracting Equation 3.A.9 from Equation 3.A.8 yields

$$e_{3x}e_{2y} - e_{2x}e_{3y} = -\sin\beta\,\sin\gamma(\cos^2\alpha + \sin^2\alpha)$$
$$= -\sin\beta\,\sin\gamma \tag{3.A.10}$$

□   □   □

$$e_{1x}e_{3y} = -\cos\alpha\,\sin\beta(\cos\alpha\,\cos\gamma - \sin\alpha\,\cos\beta\,\sin\gamma)$$
$$= -\sin\beta\,\cos\gamma\,\sin^2\alpha + \sin\alpha\,\sin\beta\,\cos\alpha\,\cos\beta\,\cos\gamma \tag{3.A.11}$$

$$e_{3x}e_{1y} = \sin\alpha\,\sin\beta(\sin\alpha\,\cos\gamma + \cos\alpha\,\cos\beta\,\sin\gamma)$$
$$= \sin\beta\,\cos\gamma\,\sin^2\alpha + \sin\alpha\,\sin\beta\,\cos\alpha\,\cos\beta\,\cos\gamma \tag{3.A.12}$$

Subtracting Equation 3.A.12 from Equation 3.A.11 yields

$$e_{1x}e_{3y} - e_{3x}e_{1y} = -\sin\beta\,\cos\gamma(\cos^2\alpha + \sin^2\alpha)$$
$$= -\sin\beta\,\cos\gamma \tag{3.A.13}$$

□   □   □

Finally dividing Equation 3.A.10 by Equation 3.A.13, we obtain

$$\frac{e_{3x}e_{2y} - e_{2x}e_{3y}}{e_{1x}e_{3y} - e_{3x}e_{1y}} = \frac{-\sin\beta\,\sin\gamma}{-\sin\beta\,\cos\gamma} = \tan\gamma \tag{3.A.14}$$

provided that $\sin\gamma$ is not equal to zero.

$$\gamma = \arctan\frac{e_{3x}e_{2y} - e_{2x}e_{3y}}{e_{1x}e_{3y} - e_{3x}e_{1y}} \tag{3.36}$$

Proof of Equations 3.38 and 3.40 is left for the reader to derive in a similar manner.

Appendix 3.B Equation of a plane

The equation of a plane is easily determined once it is recognized that a plane S is specified when one point on it is known together with any vector perpendicular to it. To derive the equation of the S plane consider Figure 3.B.1.

Let r be the position vector relative to the origin O of a point P on the S plane and n be a vector normal from the origin O so that the corresponding unit

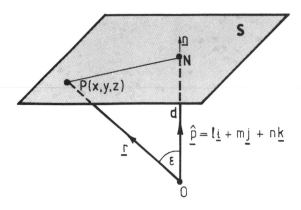

Fig. 3.B.1   Vector equation of a plane $r \circ \hat{n} = d$.

normal is $\hat{n} = n/|n|$. Further, let the perpendicular distance $\overline{ON}$ be d. Then for any point on the plane we have $\overline{OP} \cos \varepsilon = d$. In terms of vectors this is

$$r \circ \hat{n} = d \tag{3.B.1}$$

This is the vector equation of the plane in standard form. If the distance d in Equation 3.B.1 is positive then the plane lies on the side of the origin towards which n is directed, otherwise it lies on the opposite side.

To express the result in Equation 3.B.1 in cartesian form let

$$r = xi + yj + zk$$

and the unit normal

$$\hat{n} = \frac{n}{|n|} = li + mj + nk$$

Equation 3.B.1 becomes

$$lx + my + nz = d \tag{3.B.2}$$

and is the standard cartesian form of the plane having for its unit normal the vector $li + mj + nk$ and lying at a perpendicular distance d from the origin. If $d = 0$ the plane passes through the origin.

Appendix 3.C

Proof of Equation 3.63
$$e_{2z}e_{3x} = \cos \theta \sin \psi (\sin \phi \sin \psi + \cos \phi \sin \theta \cos \psi)$$

$e_{3z}e_{2x} = \cos\theta\,\cos\psi(-\sin\phi\,\cos\psi + \cos\phi\,\sin\theta\,\sin\psi)$

$e_{2z}e_{3x} - e_{3z}e_{2x} = \sin\phi\,\cos\theta$

$e_{3z}e_{2y} = \cos\theta\,\cos\psi(\cos\phi\,\cos\psi + \sin\phi\,\sin\theta\,\sin\psi)$

$e_{2z}e_{3y} = \cos\theta\,\sin\psi(-\cos\phi\,\sin\psi + \sin\phi\,\sin\theta\,\cos\psi)$

$e_{3z}e_{2y} - e_{2z}e_{3y} = \cos\phi\,\cos\theta$

Proof of Equation 3.64

$e_{1y}e_{3x} = \sin\phi\,\cos\psi(\sin\phi\,\sin\psi + \cos\phi\,\sin\theta\,\cos\psi)$

$e_{1x}e_{3y} = \cos\phi\,\cos\theta(-\cos\phi\,\sin\psi + \sin\phi\,\sin\theta\,\cos\psi)$

$e_{1y}e_{3x} - e_{1x}e_{3y} = \cos\theta\,\sin\psi$

$e_{1x}e_{2y} = \cos\phi\,\cos\theta(\cos\phi\,\cos\psi + \sin\phi\,\sin\theta\,\sin\psi)$

$e_{1y}e_{2x} = \sin\phi\,\cos\theta(-\sin\phi\,\cos\psi + \cos\phi\,\sin\theta\,\sin\psi)$

$e_{1x}e_{2y} - e_{1y}e_{2x} = \cos\theta\,\cos\psi$

3.12 REFERENCES

[3.1] Horn, B. K. P. and Inoue, H. Kinematics of the MIT-AI-Vicarm manipulator, *Working Paper 69*, Massachusetts Inst. of Technology, Artificial Intelligence Lab., May 1974.

[3.2] Jeffrey, A. *Basic Mathematics for Engineers and Technologists*, Nelson, 1974.

[3.3] Paul, R., Luh, J., Anderson, T., Benler, J., Berg, E., Brown, R., Lin, C. S., Remington, M. and Wu, C-H. Advanced industrial robot control systems, *Third Report*, Purdue University, TR-EE 80-28, July 1980.

4

ROBOT MANIPULATION TASKS

4.1   OBJECT MANIPULATION IN THE ROBOT WORKSPACE

After having studied the basic mathematical background of robot manipulation, we can now attack the problems of modeling the manipulation tasks to develop program-controlled robot operation.

The primary objective of controlling an assembly robot is to carry out manipulation tasks in its workspace. The manipulation task is described as a sequence of positions and orientations of the end effector. From the user's point of view the changing position and orientation of the workpiece, rather than the robot motion, is of paramount importance. Considering the fact that only the robot is controllable, we have to define the relationship between the robot motion and the workpiece manipulation.

To describe the manipulation tasks, the homogeneous coordinate transformation outlined in Sections 2.1 through 2.6 will be applied. Let us deal with a typical assembly process, for instance a pick-and-place procedure. Consider, for example, picking up cuboids from given points on the workbench, and inserting them into a frame lying on the same bench in a specified position. The set-up of a workplace is shown in the photograph in Figure 4.1.

The aim now is to describe the required motion of the workpieces in mathematical terms for computational purposes. We have to decide how such motion can be described in a computer language. The options for choosing a programming scheme are outlined below.

(i) At the high level of task description we specify the initial and target positions and expect the task planning program in the computer controller to find the solutions for motion of a specified robot manipulator [4.1], [4.2]. The planner will decide whether it is possible to execute that task just by using a single sequence procedure, like

GRASP→MOVE→RELEASE

or whether it is necessary to put down the manipulated object in an intermediate position somewhere in the work area. This intermediate step would be introduced for the purpose of adjusting the relative position between the end effector and the manipulated object before placing it in its final position.

To introduce such an intermediate step let us consider the single sequence programming procedure

RELEASE→CHANGE GRASP→MOVE

Figure 4.1 Initial setting for pick-and-place motion. Robotics and Mechatronics Laboratory, University of Salford.

By joining these two procedures we obtain a more applicable sequence procedure as follows

GRASP→MOVE→RELEASE→
CHANGE GRASP→MOVE→RELEASE

The GRASP→MOVE→RELEASE method is not always applicable but the joint procedure method above can always provide a solution for this type of task. Instead of considering further aspects of the general formulation of these problems, we will introduce a case study in Section 4.2 showing the implementation of a practical problem using both schemes in detail.

(ii) A somewhat more detailed high level form of task description is involved when a robot program deals not only with the initial and final states but also with the actions leading from one situation to another. Such a form of description can be called high level if in the task description program we refer to the objects to be manipulated by their names and not by the explicit grasp points [4.3]. So using such a method we could tell the planner to grasp the object, move it to a given point in the workspace, release it, depart from it, approach it from another side, grasp it again, move it to the target position and release it again. We consider it high level because the actual grasp positions and directions have to be inferred from a geometrical database.

When the objects to be manipulated are of simple shape the controlling computer can readily choose the manipulator states convenient for grasp and release. The change from one state to another can then be performed by a transformation such as that given in Equation 4.3 on an object description matrix such as those provided in Equations 4.1 and 4.2. The procedure results in a sequence of spatial points and orientations that the manipulator has to pass through when performing the task. The scheme, as implemented for robot control, is well presented by Taylor [4.4].

(iii) A widely used form of task description is obtainable by programming the robot in terms of cartesian coordinates relative to the hand or the tool center point. Spatial points to be passed through are defined either by explicitly specifying their cartesian coordinates or by guiding the robot through the points and giving names for the locations. Points to be traversed by the robot are later referred to by their names. This level of task programming is widely used in industrial assembly robot control systems such as VAL of the Unimation PUMA system [4.5].

(iv) At the lower programming level, the operator simply guides the robot through the series of desired gripper states and the joint positions at each desired cartesian point are recorded and can be played back at request. In this case playing back the complete sequence of the joints' set values constitutes the performance of the robot's task. This level of programming lacks kinematic computation.

4.2   *ILLUSTRATIVE EXAMPLE 8*
DESIGN OF MOTION FOR PICK-AND-PLACE MANIPULATION

Specification

The robot system application which we describe here is to pick up a cuboid from a given position on the workbench and place it into a frame lying on the same desk in a fixed state. The dimensions of both cuboid and frame are shown in Figures 4.2(*a*) and 4.2(*b*). The initial states of the objects and the robot are shown in Figure 4.2(*c*) and the target state inside the frame in Figure 4.2(*d*). The aim of the exercise is to show the use of homogeneous transformations for both processes:

(1) GRASP→MOVE→RELEASE
(2) GRASP→MOVE→RELEASE→
      CHANGE GRASP→MOVE→RELEASE

Design objectives

(i) Derive the mathematical description of both objects, i.e. cuboid and frame, in matrix form. The numbers in Figures 4.2(*a*) and 4.2(*b*) refer to the vertices of the cuboid and the frame and indicate the sequence of the vertices to proceed.

(ii) Derive for a GRASP→MOVE→RELEASE motion the mathematical description of the cuboid in its initial state as shown in Figure 4.2(*c*). By using the initial and target transformation matrices $H_{0,i}$ and $H_{0,t}$, we transfer the cuboid by positioning its reference origin into the nearest point to the origin of the robot's base coordinate system inside the frame, i.e. to point T shown in the plan view in Figure 4.2(*e*). It is also required to demonstrate any procedure for deriving the $H_{i,t}$ matrix between the initial and target position. The dimensions of the target position inside the frame can be calculated from both Figures 4.2(*c*) and 4.2(*e*).

(iii) Derive for a GRASP→MOVE→RELEASE→CHANGE GRASP→MOVE→RELEASE motion the mathematical description of the cuboid in an intermediate position. Apply the same specification as in (ii) for your calculation. The initial and intermediate origins O_i and O_m remain identical as far as there is no translatory motion. The suggested

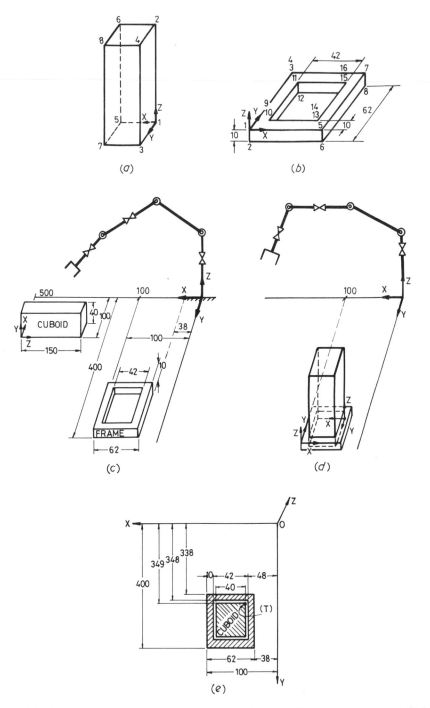

Figure 4.2   Pick-and-place manipulation with cuboid and frame assembly. (*a*) Cuboid and its coordinate system. (*b*) Frame and its coordinate system. (*c*) Cuboid and frame in the workspace. (*d*) Cuboid and frame in the task position. (*e*) Plan view of Figure 4.2(*d*).

intermediate position is shown in Figure 4.4. The position of the target point T remains unchanged.

(iv) Derive the transformation matrix of the cuboid relative to the frame. Firstly rotate the assembled cuboid-frame workpiece 90° about the z axis and then move it to a new position given by the following coordinates with respect to the base frame:

$$x = 100 \text{ units}$$
$$y = 200 \text{ units}$$
$$z = 10 \text{ units}$$

Solutions

(i) Note that a coordinate system is fixed to both cuboid and frame and their matrix descriptions are given in Equations 4.1 and 4.2 respectively:

$$[\text{cuboid}] = \begin{bmatrix} 0 & 0 & 0 & 0 & 40 & 40 & 40 & 40 \\ 0 & 0 & 40 & 40 & 0 & 0 & 40 & 40 \\ 0 & 150 & 0 & 150 & 0 & 150 & 0 & 150 \\ 1 & 1 & 1 & 1 & 1 & 1 & 1 & 1 \end{bmatrix} \tag{4.1}$$

In Equation 4.1 one of the cuboid's vertices has been chosen to be the origin which will be the reference point to the tool coordinate system (see vertex numbered 1 in Figure 4.2(a)).

$$[\text{frame}] =$$

$$\begin{bmatrix} 0 & 0 & 0 & 0 & 62 & 62 & 62 & 62 & 10 & 10 & 10 & 10 & 52 & 52 & 52 & 52 \\ 0 & 0 & 62 & 62 & 0 & 0 & 62 & 62 & 10 & 10 & 52 & 52 & 10 & 10 & 52 & 52 \\ 0 & -10 & 0 & -10 & 0 & -10 & 0 & -10 & 0 & -10 & 0 & -10 & 0 & -10 & 0 & -10 \\ 1 & 1 & 1 & 1 & 1 & 1 & 1 & 1 & 1 & 1 & 1 & 1 & 1 & 1 & 1 & 1 \end{bmatrix}$$

$$\tag{4.2}$$

In Equation 4.2 one of the frame's vertices represents the origin which will be the reference point to the tool coordinate system (see vertex numbered 1 in Figure 4.2(b).

Note that the columns of both the cuboid and frame matrices represent the coordinates of their vertices in their own reference coordinate system (see Figures 4.2(a) and 4.2(b)).

(ii) The steps of the procedure for the GRASP→MOVE→RELEASE

motion is graphically shown in Figure 4.3. The subscripts i and t stand for the initial and target positions respectively.

For the initial state of the cuboid the homogeneous transformation matrix $H_{0,i}$ is derived in numerical terms from Figure 4.2(*c*) by using Equation 2.5.

$$H_{0,i} = \begin{bmatrix} 0 & 0 & -1 & 500 \\ -1 & 0 & 0 & 100 \\ 0 & 1 & 0 & 0 \\ 0 & 0 & 0 & 1 \end{bmatrix} \tag{4.3}$$

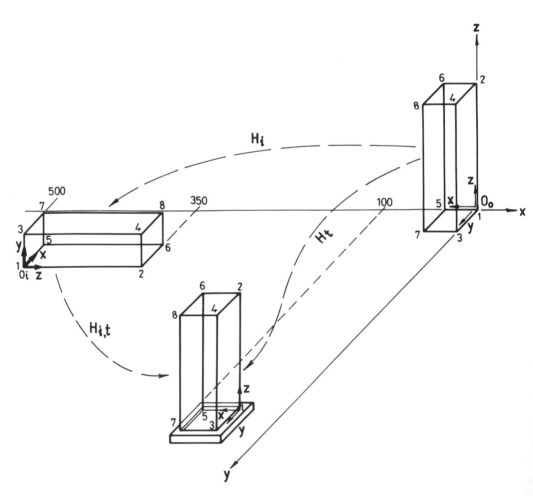

Figure 4.3   Steps of the homogeneous transformation for the GRASP→MOVE→ RELEASE motion.

It can be seen that the first three columns of the transformation matrix $H_{0,i}$ determine the direction of axes of the rotated coordinate system, while the fourth column gives the position of the reference origin of the cuboid. The initial state of the cuboid is defined by the product of the transformation matrix $H_{0,i}$ and the object matrix [cuboid]:

$$[\text{cuboid}]_i = H_{0,i} * [\text{cuboid}] \qquad (4.4)$$

The matrix [cuboid] stands with no subscript which always means that it is referred to the base coordinate frame. The initial state of the cuboid in matrix terms becomes

[cuboid]$_i$ =

$$
\begin{bmatrix}
0 & 0 & -1 & 500 \\
-1 & 0 & 0 & 100 \\
0 & 1 & 0 & 0 \\
0 & 0 & 0 & 1
\end{bmatrix}
*
\begin{bmatrix}
0 & 0 & 0 & 0 & 40 & 40 & 40 & 40 \\
0 & 0 & 40 & 40 & 0 & 0 & 40 & 40 \\
0 & 150 & 0 & 150 & 0 & 150 & 0 & 150 \\
1 & 1 & 1 & 1 & 1 & 1 & 1 & 1
\end{bmatrix}
\qquad (4.5)
$$

The product of these two matrices is

$$
[\text{cuboid}]_i =
\begin{bmatrix}
500 & 350 & 500 & 350 & 500 & 350 & 500 & 350 \\
100 & 100 & 100 & 100 & 60 & 60 & 60 & 60 \\
0 & 0 & 40 & 40 & 0 & 0 & 40 & 40 \\
1 & 1 & 1 & 1 & 1 & 1 & 1 & 1
\end{bmatrix}
\qquad (4.6)
$$

Let us now place the cuboid into the frame. The target position of the cuboid is defined by introducing a further transformation, namely

$$[\text{cuboid}]_t = H_{0,t} * [\text{cuboid}] \qquad (4.7)$$

The subscript t stands for the target position and the transformation matrix $H_{0,t}$ is derived in matrix form according to Figure 4.2(e) as

$$
H_{0,t} =
\begin{bmatrix}
1 & 0 & 0 & 49 \\
0 & 1 & 0 & 349 \\
0 & 0 & 1 & 0 \\
0 & 0 & 0 & 1
\end{bmatrix}
\qquad (4.8)
$$

The position of the cuboid in the frame is described by the matrix $[\text{cuboid}]_{t}$, namely

$$[\text{cuboid}]_t =$$

$$\begin{bmatrix} 1 & 0 & 0 & 49 \\ 0 & 1 & 0 & 349 \\ 0 & 0 & 1 & 0 \\ 0 & 0 & 0 & 1 \end{bmatrix} * \begin{bmatrix} 0 & 0 & 0 & 0 & 40 & 40 & 40 & 40 \\ 0 & 0 & 40 & 40 & 0 & 0 & 40 & 40 \\ 0 & 150 & 0 & 150 & 0 & 150 & 0 & 150 \\ 1 & 1 & 1 & 1 & 1 & 1 & 1 & 1 \end{bmatrix} \tag{4.9}$$

The product of these two matrices is

$$[\text{cuboid}]_t = \begin{bmatrix} 49 & 49 & 49 & 49 & 89 & 89 & 89 & 89 \\ 349 & 349 & 389 & 389 & 349 & 349 & 389 & 389 \\ 0 & 150 & 0 & 150 & 0 & 150 & 0 & 150 \\ 1 & 1 & 1 & 1 & 1 & 1 & 1 & 1 \end{bmatrix} \tag{4.10}$$

The matrix $[\text{cuboid}]_t$ describes the position and orientation of the cuboid in its new position inside the frame shown in Figure 4.2(d). The transformation between the initial and target states is defined by the matrix $H_{i,t}$. The steps of the transformation are shown in Figure 4.3. Since the initial matrix $H_{0,1}$ and the target matrix $H_{0,t}$ are already available, the transformation matrix between the initial and target positions is readily obtainable by matrix manipulation as follows

$$H_{0,t} = H_{0,i} * H_{i,t}$$

$$H_{i,t} = H_{0,i}^{-1} * H_{0,t} \tag{4.11}$$

$$H_{i,t} = \begin{bmatrix} 0 & -1 & 0 & 100 \\ 0 & 0 & 1 & 0 \\ -1 & 0 & 0 & 500 \\ 0 & 0 & 0 & 1 \end{bmatrix} * \begin{bmatrix} 1 & 0 & 0 & 49 \\ 0 & 1 & 0 & 349 \\ 0 & 0 & 1 & 0 \\ 0 & 0 & 0 & 1 \end{bmatrix}$$

$$= \begin{bmatrix} 0 & -1 & 0 & -249 \\ 0 & 0 & 1 & 0 \\ -1 & 0 & 0 & 451 \\ 0 & 0 & 0 & 1 \end{bmatrix} \tag{4.12}$$

The $H_{i,t}$ matrix represents the transformation to the target position at T with respect to the coordinate system of the cuboid's initial position. Alternatively the $H_{i,t}$ can be directly read out from Figures 4.2(c) and 4.2(d) and the product of $H_{i,t}$ and [cuboid]$_i$ matrices again results in the state of the cuboid in the target position.

Finally we note that the handling of this procedure constitutes a short cut version of the GRASP→MOVE→RELEASE routine for a pick-and-place motion procedure. This method is not always feasible or possible in practice.

(iii) For the intermediate state of the cuboid, PUT DOWN→RELEASE →CHANGE GRASP shown in Figure 4.4, the reference point has been chosen as if it was in its initial position. Similarly to Equation 4.4 the intermediate state

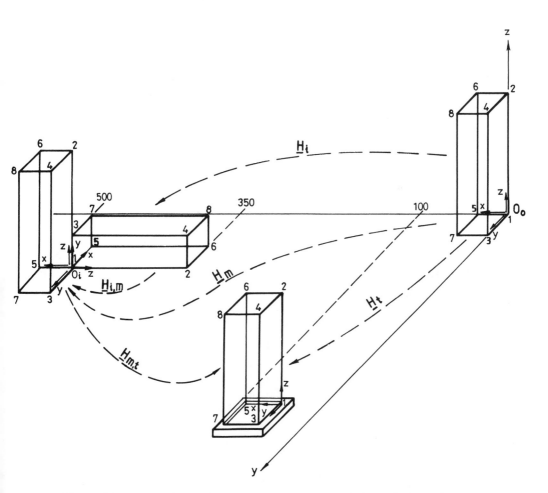

Figure 4.4 Steps of the homogeneous transformation positioning the intermediate state for the PUT DOWN→RELEASE→CHANGE GRASP procedure.

of the cuboid is expressed as

$$[\text{cuboid}]_m = H_{i,m} * [\text{cuboid}]_i \tag{4.13}$$

where the subscript m stands for the intermediate state of the cuboid.

The new coordinate axes in the direction of the new unitary vectors are given as

$$
\begin{aligned}
x \text{ axis is } [\;\; 0 \quad\;\; 0 \quad -1]^T \\
y \text{ axis is } [-1 \quad\;\; 0 \quad\;\; 0]^T \\
z \text{ axis is } [\;\; 0 \quad\;\; 1 \quad\;\; 0]^T
\end{aligned} \tag{4.14a}
$$

and k vector is $[\;\; 0 \quad\;\; 0 \quad\;\; 0]^T$

The first three column vectors constitute the rotation submatrix in the transformation matrix from the cuboid's initial state to its intermediate state. Note that there is no displacement of the origin O_i.

The initial–intermediate transformation matrix $H_{i,m}$ is

$$
H_{i,m} = \begin{bmatrix}
0 & -1 & 0 & 0 \\
0 & 0 & 1 & 0 \\
-1 & 0 & 0 & 0 \\
0 & 0 & 0 & 1
\end{bmatrix} \tag{4.14b}
$$

Thus the intermediate transformation matrix is

$$H_{0,m} = H_{0,i} * H_{i,m} \tag{4.15a}$$

Substituting for $H_{0,i}$ from Equation 4.3 and $H_{i,m}$ from Equation 4.14b the matrix $H_{0,m}$ in expanded form becomes

$$
H_{0,m} = \begin{bmatrix}
0 & 0 & -1 & 500 \\
-1 & 0 & 1 & 100 \\
0 & 1 & 0 & 0 \\
0 & 0 & 0 & 1
\end{bmatrix} * \begin{bmatrix}
0 & -1 & 0 & 0 \\
0 & 0 & 1 & 0 \\
-1 & 0 & 0 & 0 \\
0 & 0 & 0 & 1
\end{bmatrix}
$$

$$
= \begin{bmatrix}
1 & 0 & 0 & 500 \\
0 & 1 & 0 & 100 \\
0 & 0 & 1 & 0 \\
0 & 0 & 0 & 1
\end{bmatrix} \tag{4.15b}
$$

The rotation submatrix in Equation 4.15b is a unit matrix and indicates that the cuboid is in the same orientation as it was before in its initial reference position. The cuboid in its intermediate state is defined as

$$[\text{cuboid}]_m = H_{0,m} * [\text{cuboid}] \tag{4.16a}$$

and multiplying matrices in Equations 4.1 and 4.15b results in the intermediate position of the cuboid as

$$[\text{cuboid}]_m = \begin{bmatrix} 500 & 500 & 500 & 500 & 540 & 540 & 540 & 540 \\ 100 & 100 & 140 & 140 & 100 & 100 & 140 & 140 \\ 0 & 150 & 0 & 150 & 0 & 150 & 0 & 150 \\ 1 & 1 & 1 & 1 & 1 & 1 & 1 & 1 \end{bmatrix} \tag{4.16b}$$

In the next step the cuboid is going to be inserted from its intermediate position into the frame as shown in Figure 4.2(*d*). Using the cuboid's intermediate state to interpret the motion, its target state is obtained as

$$[\text{cuboid}]_t = H_{m,t} * [\text{cuboid}]_m \tag{4.17}$$

where $H_{m,t}$ is the transformation matrix for the motion between the intermediate and target positions and its derivation is resolved as

$$H_{0,t} = H_{0,m} * H_{m,t}$$
$$H_{m,t} = H_{0,m}^{-1} * H_{0,t} \tag{4.18}$$

There is no need, however, to calculate either the inverse of the matrix $H_{0,m}$ or to carry out the matrix multiplication since there is only a turn in the (x, y) plane between the intermediate and target states.

$$H_{m,t} = \begin{bmatrix} 1 & 0 & 0 & -451 \\ 0 & 1 & 1 & 249 \\ 0 & 0 & 1 & 0 \\ 0 & 0 & 0 & 1 \end{bmatrix} \tag{4.19}$$

Equation 4.17 in expanded form results in Equation 4.20, which is the same as

Equation 4.19, namely

$[\text{cuboid}]_t$

$$
= \begin{bmatrix} 1 & 0 & 0 & -451 \\ 0 & 1 & 0 & 249 \\ 0 & 0 & 1 & 0 \\ 0 & 0 & 0 & 1 \end{bmatrix} * \begin{bmatrix} 500 & 500 & 500 & 500 & 540 & 540 & 540 & 540 \\ 100 & 100 & 140 & 140 & 100 & 100 & 140 & 140 \\ 0 & 150 & 0 & 150 & 0 & 150 & 0 & 150 \\ 1 & 1 & 1 & 1 & 1 & 1 & 1 & 1 \end{bmatrix}
$$

$$
= \begin{bmatrix} 49 & 49 & 49 & 49 & 89 & 89 & 89 & 89 \\ 349 & 349 & 389 & 389 & 349 & 349 & 389 & 389 \\ 0 & 150 & 0 & 150 & 0 & 150 & 0 & 150 \\ 1 & 1 & 1 & 1 & 1 & 1 & 1 & 1 \end{bmatrix} \tag{4.20}
$$

In this interpretation the manipulation is considered as a sequence of homogeneous transformations. It is worth mentioning that Equations 4.10 and 4.20 are equal as they have to be.

(iv) In order to make the pick-and-place procedure more understandable and to ease its mathematical description, computation and, later on, the robot programming, we will demonstrate the use of the relative transformation which was outlined in Section 2.3 and described in further detail in Section 4.4.

The question now is how to perform a simple transformation procedure mathematically for moving the workpiece in the frame and then how to derive the coordinates of the workpiece in the base coordinate system, i.e.

$$
[\text{workpiece}]_{\text{base coordinate system}} = ? \tag{4.21}
$$

In our case the relative transformation procedure for the assembled cuboid/frame workpiece appears in matrix presentation as

$$
[\text{cuboid}]_{\text{base}} = [\text{frame}]_{\text{base}} * [\text{cuboid}]_{\text{frame}} \tag{4.22}
$$

$$
H_{cb} = H_{fb} * H_{cf}
$$

The relative position of the cuboid to the frame is shown in the target position of the cuboid in Figure 4.2(d). The transformation below constitutes the relative state of the cuboid with respect to the frame

$$H_{cf} = [\text{cuboid}]_{\text{frame}} = \begin{bmatrix} -1 & 0 & 0 & 51 \\ 0 & -1 & 0 & 51 \\ 0 & 0 & 1 & -10 \\ 0 & 0 & 0 & 1 \end{bmatrix} \qquad (4.23)$$

Let us now perform a transformation with respect to the coordinate system containing the cuboid and then position it at a point with respect to the base given in the specification.

$$H_{fb} = [\text{frame}]_{\text{base}} = \begin{bmatrix} 0 & -1 & 0 & 100 \\ 1 & 0 & 0 & 400 \\ 0 & 0 & 1 & 10 \\ 0 & 0 & 0 & 1 \end{bmatrix} \qquad (4.24)$$

The cuboid moves together with the frame and of course keeps its state with respect to the frame unchanged. Thus, the state of the cuboid in the base coordinate system can be described by a matrix representation as complies with Equation 4.22.

$$H_{cb} = [\text{cuboid}]_{\text{base}}$$

$$= \begin{bmatrix} 0 & -1 & 0 & 100 \\ 1 & 0 & 0 & 400 \\ 0 & 0 & 1 & 10 \\ 0 & 0 & 0 & 1 \end{bmatrix} * \begin{bmatrix} -1 & 0 & 0 & 51 \\ 0 & -1 & 0 & 51 \\ 0 & 0 & 1 & -10 \\ 0 & 0 & 0 & 1 \end{bmatrix}$$

$$= \begin{bmatrix} 0 & 1 & 0 & 51 \\ -1 & 0 & 0 & 451 \\ 0 & 0 & 1 & 0 \\ 0 & 0 & 0 & 1 \end{bmatrix} \qquad (4.25)$$

4.3   TASK FORMULATION

We have seen what an assembly task looks like from the point of view of the objects taking part in it. In this section we are going to discuss the same task with respect to high level manipulator control. The task, e.g. picking up a cuboid and inserting it into a frame, will now be formulated in more detail. In order to do this we have to define a series of robot gripper states (positions and orientations) which we will denote by S(i) where $i = 1,2,3, \ldots$ stands for the gripper states, shown in sequential steps, in Figure 4.5. We will now describe the pick-and-place task as a series of MOVE, GRASP and RELEASE instructions.

(1)	MOVE S(1)	go over cuboid
(2)	MOVE S(2)	go down to cuboid
(3)	GRASP	
(4)	MOVE S(3)	lift the cuboid
(5)	MOVE S(4)	move over intermediate position change orientation of the cuboid
(6)	MOVE S(5)	go down to the table with cuboid
(7)	RELEASE	
(8)	MOVE S(6)	pull back the gripper
(9)	MOVE S(7)	go over cuboid while changing gripper orientation
(10)	MOVE S(8)	go down to cuboid
(11)	GRASP	
(12)	MOVE S(9)	lift it
(13)	MOVE S(10)	go over frame
(14)	MOVE S(11)	insert cuboid into frame
(15)	RELEASE	
(16)	MOVE S(12)	pull back gripper

Any industrial robot commercially available which has the necessary number of degrees of freedom, usually 4, 5 or 6, and the required reach, can perform the program steps above. The problem now is what is the relationship of such a program to the series of homogeneous transformations characterizing the position and orientation changes of the object in the workspace. In order to find this relationship we must first specify the general geometrical structure of the manipulation tasks.

4.4   RELATIVE TRANSFORMATION IN THE ROBOT WORKSPACE

The general task of the manipulator is to carry the hand gripper or the tool from its initial position and orientation, through a number of given spatial positions and orientations, to a target location. The state of gripper/tool assembly is represented by the position of its reference point, which is called the tool center point abbreviated to TCP, and by the orientation of the axis of the

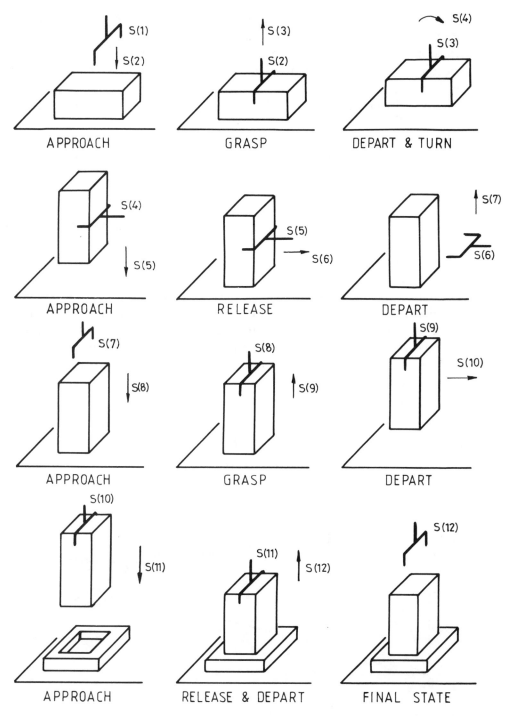

Figure 4.5 States of the cuboid in a pick-and-place process in sequential order.

manipulator's last link, which is usually a roll and becomes the tool z axis. In order to support the interpretation of the derivation of the relative transformation in Equation 4.26, consult Figure 4.6. Note that both the world coordinate and the actual workplace coordinate systems are fixed to the horizontal plane.

Any gripper state identified by a state matrix $S(i)$ can be reached in two ways as follows:

(1) From the manipulator's point of view the gripper's state is expressed by

$$S(i) = B * T_6(i) * E \tag{4.26}$$

(2) From the task's point of view the ith state of the gripper is defined by

$$S(i) = W * G(i) \tag{4.27}$$

The matrix B in Figure 4.6 designates the state of the manipulator's base in the world, i.e. in the reference coordinate system. If the manipulator is fixed, as it is in our case, matrix B is constant. For the sake of simplicity we shall consider $B = I$, i.e. B equals the unit matrix which means that the world reference and the manipulator base reference systems are identical. Thus, Equations 4.26 and 4.27 are equal and reduced to

$$T_6(i) * E = W * G(i) \tag{4.28}$$

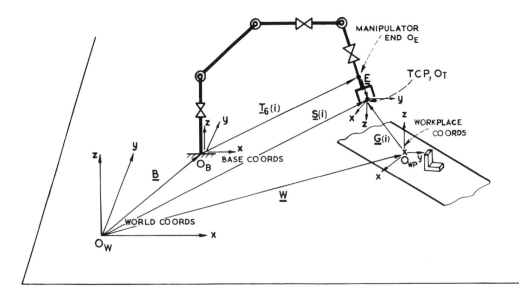

Figure 4.6 Relative transformation in the robot workspace.

The matrix $T_6(i)$ represents the state of the end effector of a six degrees of freedom manipulator, with respect to its own base coordinate system of origin O_B, in the ith step along its trajectory. This can be computed from the actual joint angles. Note that the use of the matrix $T_6(i)$ in robot technology is already a convention throughout the robotics literature, therefore the subscript 6 is also bold and the matrix often appears without notifying the steps, that is as T_6.

The matrix E represents the state of the end effector, that is the TCP with respect to the end of the manipulator. While moving from the initial state to the target, the E matrix remains constant. However, for the purpose of different tasks or partial tasks within the main task, the manipulator may change tools, which can also change the values in matrix E.

The matrix W represents the workplace, that is the actual coordinate system for the manipulation, which can be fixed to a machine tool, an assembly, a measuring or testing device, a conveyor belt or a pallet box, etc.

The matrix $G(i)$ is the transformation matrix for grasping or assembling which defines the desired state, either of the gripper or tool, with respect to the actual coordinate system of the workplace, in the ith step, within the manipulation sequence.

Whichever way the manipulator task is defined before its execution, all gripper states to be passed through must be defined. Knowing these states and the actual values of matrix E, the corresponding sequence of transformations for $T_6(i)$ must be created, where $i = 1, 2, ..., n$ indicates the states to be passed through. Section 4.5 will now outline the calculation of the matrix $T_6(i)$ for further applications developed in Illustrative example 9.

4.5 FURTHER USES OF RELATIVE TRANSFORMATIONS

In Section 4.2 we dealt with just one workpiece to be manipulated in the workspace of the robot at any one time. Let us now extend this experiment to that shown in Figures 4.1 and 4.7. There are several workpieces, for instance suppose there are three cuboids and one frame with three openings. This set up creates three cycle loops for this pick-and-place process. The state of the cuboids with respect to the world (reference) coordinate system is defined by $C[k]$ and the state of the frame openings with respect to the frame (reference) coordinate system is provided by $O_F[j]$, where k is the number of cuboids and j is the number of openings in the frame as shown in Figure 4.7. The subscript F denotes the frame.

In this experiment, while executing a cycle of the inserting task, the TCP passes through the states S(1)–S(4). We note that the cuboids this time will be inserted without changing the gripper. The mathematical representation of the task requires us to prepare a series of coordinate transformations. The lower index of the transformation matrices refers to the reference coordinate frames,

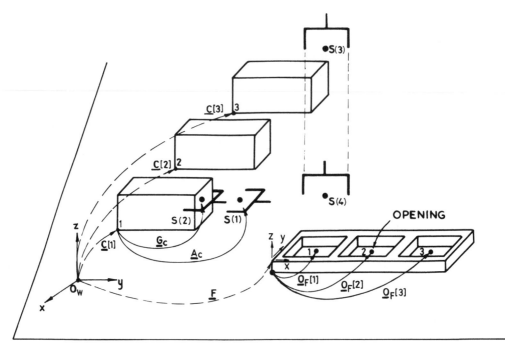

Figure 4.7   Cuboids in the workarea.

and the matrices without index are defined in the base (world) coordinate system of origin O_B.

The mathematical notation used in the task procedure is defined below (see Figures 4.7 to 4.10):

$C[k]$ is the matrix representing the state of the kth cuboid in the world (reference) coordinate system.

A_C is the state of the gripper when it is approaching the cuboid; the approach state is defined with respect to the cuboid.

G_C is the state of the gripper when it is grasping the cuboid; the grasping state is defined with respect to each cuboid.

F is the matrix representing the state of the frame with respect to the world coordinate system.

$O_F[j]$ is the state of the jth opening in the coordinate system of the frame.

CA_O is the state of the cuboid when it is approaching the frame; the approach state is defined with respect to the frame opening shown in Figure 4.10(a).

CF_O is the state of the cuboid with respect to the frame opening when it is inserted into the frame shown in Figure 4.10(b).

The task can be represented by a set of transformation equations and be solved one by one for $T_6(i)$. Thus a series of commands for the manipulator is required to be created. The equations describing the current cycling process are developed as follows (see Figure 4.6):

$$
\begin{aligned}
\text{S(1)} \quad & T_6(1) * E = C[k] * A_C \\
\text{S(2)} \quad & T_6(2) * E = C[k] * G_C \\
\text{S(3)} \quad & T_6(3) * E = F * O_F[j] * CA_O * G_C \\
\text{S(4)} \quad & T_6(4) * E = F * O_F[j] * CF_O * G_C
\end{aligned}
\tag{4.29}
$$

When executing a practical task, the gripper of the manipulator may pass a greater number of explicitly defined points than outlined in Equations 4.29. In order to demonstrate the structure of such tasks, this simplified version of the description of the process in Equations 4.29 is sufficient to illustrate the mathematical process. The advantage of the structural description is that when placing another cuboid into another opening, only the $C[k]$ and $O_F[j]$ transformations have new values, therefore we can execute another cycle of the task under the control of the same program as before. Because the cuboids may be transported on a conveyor belt, the values, of $C[k]$ can probably be the same since the cuboids are always picked up from the same point of the conveyor belt. Of course different $C[k]$ values apply for a pallet initially containing the cuboids. Transformations appearing on the right-hand side of the set of Equations 4.29, i.e. transformations producing the desired manipulator states, can be obtained from several sources as outlined below:

(1) The values of $O_F[j]$, i.e. the coordinates of the openings inside the frame, can for instance be retrieved from the engineering drawing of the parts and typed in explicitly or in the case of CAD/CAM automatic planning from a geometrical data base (see Section 4.11). A geometrical data base may also contain information referring to the way of grasping the given object, i.e. to the information referring to the A_C and G_C transformations in Equations 4.29.

(2) Instead of geometrical inference systems we can obtain certain geometrical parameters by using visual inputs. In the above example a camera looking at the scene could fix the position and orientation of the cuboid. This method of acquiring information is typical in intelligent assembly systems [4.6]. Visually acquired information is frequently used to obtain the state of objects, the position and/or orientation of which is not known beforehand or is changing during the execution of the task (see Section 2.5). In our case, matrices $C[k]$ and F could be obtained in this way.

(3) In the case of a conveyor belt, a vibratory feeder or similar fixed feeding

equipment, the picking up of the arriving workpieces takes place at a given point and usually in a fixed orientation.

(4) The usual way to specify transformations like A_C and G_C, if they are not inferred from the geometric model, is that the manipulator is guided by a human operator to these points. The values of $T_6(i)$ are evaluated by measuring the joint angles of the arm and by solving the set of transformation equations for the desired unknowns. This process is called teaching by showing or teaching by leading through.

4.6   *ILLUSTRATIVE EXAMPLE 9*
DESIGN OF TOOL MANIPULATION IN THE ROBOT
WORKSPACE

Specification

In this exercise we will deal with the transformations that appeared in Equations 4.29. The origin of the world coordinate system will be the same as that of the manipulator base frame. Thus the matrix B in Equation 4.26 is equal

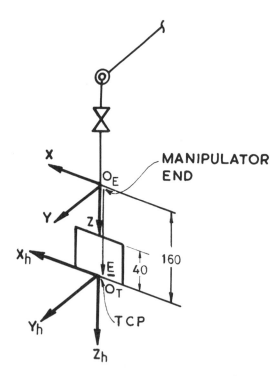

Figure 4.8   Tool coordinate frame. The TCP is at the origin of the tool coordinate frame O_T.

to unit matrix I constituting the unit transformation. The state of the gripper is defined with respect to the manipulator end O_E (see Figure 4.8). The distance between the manipulator end and the TCP is 160 mm. The dimension of the cuboid is given as 40 mm × 40 mm × 150 mm (see Figure 4.2(c)). The dimension of the frame containing three square openings is shown in Figure 4.9(a). The initial setup of the work area with the distances from the reference coordinate system to the cuboids and frame is depicted in Figure 4.9(b).

Design objectives

(i) Derive the matrix E in Equations 4.29 in the tool coordinate frame at the manipulator wrist flange (O_E) (see Figure 4.8).
(ii) Derive the matrices $O_F[j]$ by homogeneous transformation for the three square openings. The origins $O_F[j]$ are on the upper surface of the openings (see Figure 4.9(b)).
(iii) Calculate the position and orientation of the manipulator end in the S(i) states before it does the final approach to each cuboid, i.e. $T_6(1)...T_6(4)$. Note that all matrices encountered in Equations 4.29 must be derived beforehand in their numerical terms.

Solutions

(i)

$$E = \begin{bmatrix} 1 & 0 & 0 & 0 \\ 0 & 1 & 0 & 0 \\ 0 & 0 & 1 & 160 \\ 0 & 0 & 0 & 1 \end{bmatrix} \tag{4.30}$$

(ii) See Figures 4.7 and 4.9(a).

$$O_F[1] = \begin{bmatrix} 0 & 1 & 0 & 31 \\ -1 & 0 & 0 & 31 \\ 0 & 0 & 1 & 0 \\ 0 & 0 & 0 & 1 \end{bmatrix} \tag{4.31}$$

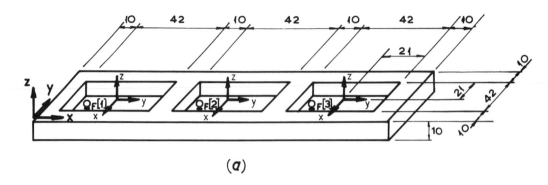

(*a*)

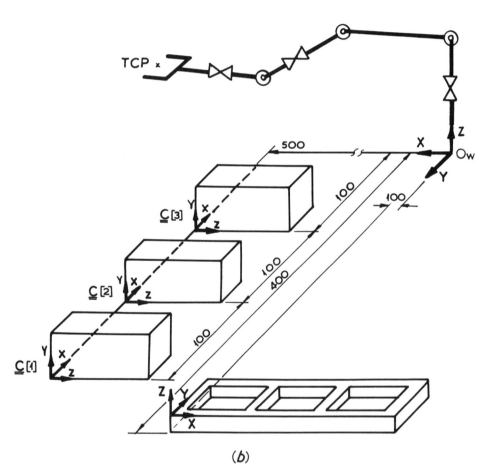

(*b*)

Figure 4.9   The initial set-up of the robot workarea. (*a*) Dimensions of the frame containing the square openings and the coordinate systems of their centers on the upper surface of the openings. (*b*) Reference coordinates on cuboids and the frame.

$$O_F[2] = \begin{bmatrix} 0 & 1 & 0 & 83 \\ -1 & 0 & 0 & 31 \\ 0 & 0 & 1 & 0 \\ 0 & 0 & 0 & 1 \end{bmatrix} \tag{4.32}$$

$$O_F[3] = \begin{bmatrix} 0 & 1 & 0 & 135 \\ -1 & 0 & 0 & 31 \\ 0 & 0 & 1 & 0 \\ 0 & 0 & 0 & 1 \end{bmatrix} \tag{4.33}$$

(iii) The state of the frame is expressed by matrix F (see Figures 4.7, 4.9(a) and 4.9(b))

$$F = \begin{bmatrix} -1 & 0 & 0 & 100 \\ 0 & -1 & 0 & 400 \\ 0 & 0 & 1 & 10 \\ 0 & 0 & 0 & 1 \end{bmatrix} \tag{4.34}$$

The initial states of the three cuboids are

$$C[1] = \begin{bmatrix} 0 & 0 & -1 & 500 \\ -1 & 0 & 0 & 300 \\ 0 & 1 & 0 & 0 \\ 0 & 0 & 0 & 1 \end{bmatrix} \quad C[2] = \begin{bmatrix} 0 & 0 & -1 & 500 \\ -1 & 0 & 0 & 200 \\ 0 & 1 & 0 & 0 \\ 0 & 0 & 0 & 1 \end{bmatrix}$$

$$C[3] = \begin{bmatrix} 0 & 0 & -1 & 500 \\ -1 & 0 & 0 & 100 \\ 0 & 1 & 0 & 0 \\ 0 & 0 & 0 & 1 \end{bmatrix} \tag{4.35}$$

The state of a cuboid is determined by aligning the corresponding axes on the cuboid and the frame. The approach state of the cuboid is a position just before it would be inserted into the corresponding opening of the frame (see Figure 4.10).

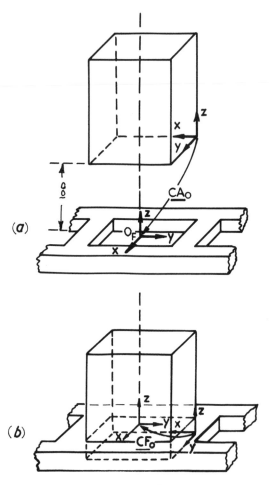

Figure 4.10 Relative positions of the cuboid to the frame. (*a*) Approach state described by CA_O matrix. (*b*) Inserted state described by CF_O matrix.

$$CA_O = \begin{bmatrix} 0 & 1 & 0 & 20 \\ -1 & 0 & 0 & 20 \\ 0 & 0 & 1 & 100 \\ 0 & 0 & 0 & 1 \end{bmatrix} \tag{4.36}$$

$$CF_O = \begin{bmatrix} 0 & 1 & 0 & 20 \\ -1 & 0 & 0 & 20 \\ 0 & 0 & 1 & -10 \\ 0 & 0 & 0 & 1 \end{bmatrix} \tag{4.37}$$

Transformations between the cuboid and the gripper are shown in Figures 4.7 and 4.11.

$$A_C = \begin{bmatrix} -1 & 0 & 0 & 20 \\ 0 & 1 & 0 & 20 \\ 0 & 0 & -1 & 250 \\ 0 & 0 & 0 & 1 \end{bmatrix} \tag{4.38}$$

$$G_C = \begin{bmatrix} -1 & 0 & 0 & 20 \\ 0 & 1 & 0 & 20 \\ 0 & 0 & -1 & 110 \\ 0 & 0 & 0 & 1 \end{bmatrix} \tag{4.39}$$

By substituting the transformation matrices from Equations 4.30 to 4.39 into the set of Equations 4.29 we get the states S(1)–S(4) through which the manipulator has to pass while executing a loop of the task. These states define the trajectory points and hold the desired trajectory.

The state of the manipulator end in the S(1), when it is approaching the cuboid $C[1]$, is presented as

$$T_6(1) = C[1] * A_C * E^{-1} \tag{4.40a}$$

$T_6(1)$

$$= \begin{bmatrix} 0 & 0 & -1 & 500 \\ -1 & 0 & 0 & 300 \\ 0 & 1 & 0 & 0 \\ 0 & 0 & 0 & 1 \end{bmatrix} * \begin{bmatrix} -1 & 0 & 0 & 20 \\ 0 & 1 & 0 & 20 \\ 0 & 0 & -1 & 250 \\ 0 & 0 & 0 & 1 \end{bmatrix} * \begin{bmatrix} 1 & 0 & 0 & 0 \\ 0 & 1 & 0 & 0 \\ 0 & 0 & 1 & -160 \\ 0 & 0 & 0 & 1 \end{bmatrix}$$

$$= \begin{bmatrix} 0 & 0 & 1 & 90 \\ 1 & 0 & 0 & 280 \\ 0 & 1 & 0 & 20 \\ 0 & 0 & 0 & 1 \end{bmatrix} \tag{4.40b}$$

In the approach state the axis of the manipulator end (z axis) must be parallel

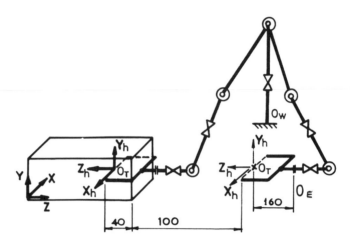

Figure 4.11   Relationship between cuboid and gripper.

with the x axis of the cuboid reference system and the normal vector of the manipulator end (y axis) must be parallel with the z axis of the cuboid coordinate system. When calculating the further trajectory points, the same method is suggested to be used as above. Then the manipulator arm will go through the states $S(1)$, $S(2)$, $S(3)$ and $S(4)$ described by the matrices

$$T_6(1) = \begin{bmatrix} 0 & 0 & 1 & 90 \\ 1 & 0 & 0 & 280 \\ 0 & 1 & 0 & 20 \\ 0 & 0 & 0 & 1 \end{bmatrix} \qquad T_6(2) = \begin{bmatrix} 0 & 0 & 1 & 230 \\ 1 & 0 & 0 & 280 \\ 0 & 1 & 0 & 20 \\ 0 & 0 & 0 & 1 \end{bmatrix}$$

$$T_6(3) = \begin{bmatrix} -1 & 0 & 0 & 69 \\ 0 & 1 & 0 & 409 \\ 0 & 0 & -1 & 380 \\ 0 & 0 & 0 & 1 \end{bmatrix} \qquad T_6(4) = \begin{bmatrix} -1 & 0 & 0 & 69 \\ 0 & 1 & 0 & 369 \\ 0 & 0 & -1 & 270 \\ 0 & 0 & 0 & 1 \end{bmatrix}$$

$$(4.41)$$

It is suggested that the reader should check the matrices $T_6(2)$, $T_6(3)$ and $T_6(4)$ which can be derived from Equations 4.29 in a way similar to that of Equation 4.40a and 4.40b. A short version of their derivation is presented in Appendix 4.A.

4.7   *ILLUSTRATIVE EXAMPLE 10*
DESIGN OF MOTION SEQUENCE FOR AN ASSEMBLY

Specification

This exercise is the continuation of Illustrative example 2 in Section 2.5. The initial set-up of the assembly parts is shown in Figure 4.12. The coordinates of the bottom center point of block B standing on the worktable are $x = 20$, $y = 20$, $z = 0$ units. The diameter of block B is 1 unit. The height of the blocks A and C is 2 units. The height of the block B is 20 units. The maximum grasp width of the

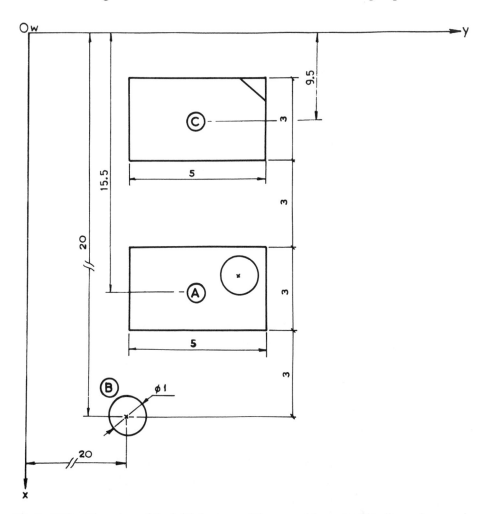

Figure 4.12   Plan view of the initial set-up of the assembly parts. The dimensions are in arbitrarily chosen units.

gripper is 4 units. The initial state of the manipulator end effector is $x=0$, $y=0$, $z=40$, $\alpha=0°$, $\beta=0°$, $\gamma=0°$, i.e. it is pointing upwards above the origin (see Figure 4.14). The approach and departure points are 5 distance units away from that face of the object whose normal vector aligns with the approach/departure direction. The TCP has to pass these points and the last link of the manipulator must align with these normal vectors when the TCP is in these points.

Design objectives

(i) Specify the states of the gripper which the TCP must pass while assembling the parts shown in Figures 2.5 and 2.6, i.e. what are the states (positions and orientations) the manipulator gripper must go through.

(ii) Specify approach and departure points for each grasping process. Use the safer gripping from side rather than from the top of the cylindrical rod (see Figure 4.13).

Solutions

(i) The solution can be summarized as follows:

(i) GRASP A
(ii) PLACE A TO O_1
(iii) GRASP B
(iv) PLACE B TO O_2
(v) GRASP C
(vi) PLACE C TO O_4

The interpretation of O_1, O_2, O_4 is self-explanatory in Figure 2.5. In this exercise a similar motion sequence has to be performed three times in order to move all three assembly components to their target place. Each motion sequence consists of the following steps:

(1) MOVE TO START–APPROACH POINT
(2) MOVE TO START–GRASP POINT
(3) GRASP
(4) MOVE TO START–DEPARTURE POINT
(5) MOVE TO TARGET–APPROACH POINT
(6) MOVE TO TARGET–RELEASE POINT
(7) RELEASE
(8) MOVE TO TARGET–DEPARTURE POINT

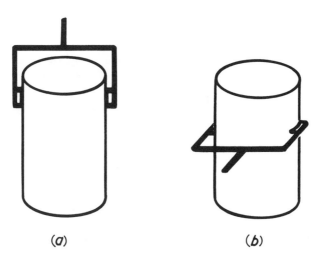

Figure 4.13 Ways of gripping. (*a*) Unsafe gripping. (*b*) Safe gripping.

The way of gripping the object and the whole assembling process is shown in Figure 4.14.

(ii) The subsequent steps comply as follows:

S(0) is the start state of the manipulator and effector.

S(1) is the approach state before grasping object A in its start position.

S(2) is the grasping manipulator state for object A in its start position.

S(3) is the departure state after grasping A. Considering the grasp width of the gripper (4 units), objects A and C can be grasped only along their longer side faces since then the required grasp width is only 3 units (see Figure 4.12). When grasping the object A we suppose the TCP to be at the center point of the upper face. By taking the dimensions of A into consideration the grasp point is $x = 15.5$, $y = 22.5$ and $z = 2$ and the last link points downwards. Thus

$$
\begin{aligned}
S(0) &= (\ 0, \quad\ \ 0, \quad 40, \quad 0°, \quad\ \ 0°, \quad 0°) \\
S(1) &= (15.5, \quad 22.5, \quad 7, \quad 0°, \quad 180°, \quad 0°) \\
S(2) &= (15.5, \quad 22.5, \quad 2, \quad 0°, \quad 180°, \quad 0°) \\
S(3) &= (15.5, \quad 22.5, \quad 7, \quad 0°, \quad 180°, \quad 0°)
\end{aligned}
$$

In the target situation the coordinates of the bottom left corner point of object A are $x = -5$, $y = 10$, $z = 10$ (see Figure 2.7).

S(4) is the approach state before releasing A in its target position.

S(5) is the releasing state.

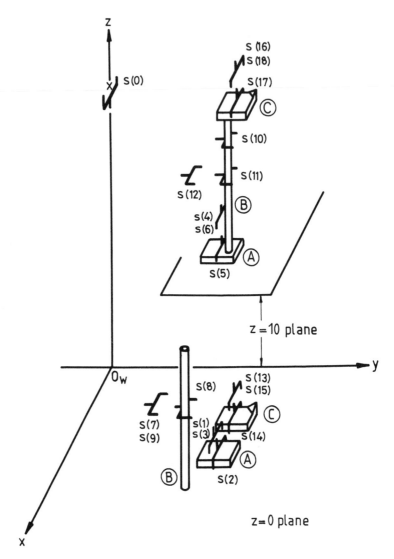

Figure 4.14   Sequential steps for assembling the parts.

S(6)   is the departure state referring to the release position of A. Thus

$$S(4) = (-6.5, \quad 12.5, \quad 17, \quad 0°, \quad 180°, \quad 0°)$$
$$S(5) = (-6.5, \quad 12.5, \quad 12, \quad 0°, \quad 180°, \quad 0°)$$
$$S(6) = (-6.5, \quad 12.5, \quad 17, \quad 0°, \quad 180°, \quad 0°)$$

Object B, the standing cylinder, can firmly be grasped by approaching it from the side. The grasping position of the TCP is $x = 20, y = 20$,

$z = 10$ units and the Euler angles are $\alpha = 0°$, $\beta = 90°$, $\gamma = 0°$ since object B is chosen to be approached from the y direction.

S(7)   is the approach state.
S(8)   is the grasping state in the start position of object B.
S(9)   is the departure state.

$$S(7) = (20, \quad 15, \quad 10, \quad 0°, \quad 90°, \quad 0°)$$
$$S(8) = (20, \quad 20, \quad 10, \quad 0°, \quad 90°, \quad 0°)$$
$$S(9) = (20, \quad 15, \quad 10, \quad 0°, \quad 90°, \quad 0°)$$

According to Figure 4.14 in the target situation the coordinates of O_2 are $x = -7$, $y = 14$, $z = 12$.

S(10) is the approach state.
S(11) is the release state for object B in the target situation.
S(12) is the departure state, which is not the same as at S(10) since the target position of B can be approached from the vertical direction while, after releasing B, departure takes place in the $-y$ direction.

$$S(10) = (-7, \quad 14, \quad 27, \quad 0°, \quad 90°, \quad 0°)$$
$$S(11) = (-7, \quad 14, \quad 22, \quad 0°, \quad 90°, \quad 0°)$$
$$S(12) = (-7, \quad 9, \quad 22, \quad 0°, \quad 90°, \quad 0°)$$

For object C the same kind of motions are required as for object A with an offset of $\Delta x = -6$, $\Delta y = 0$, $\Delta z = 0$ in the start situation and $\Delta x = -1$, $\Delta y = 2$, $\Delta z = 12$ in the target situation (see Figure 4.14).
S(13) is the approach state.
S(14) is the grasping state in the start situation.
S(15) is the departure state.
S(16) is the approach state.
S(17) is the release state in the target situation for object B.
S(18) is the departure state.

$$S(13) = (9.5, \quad 22.5, \quad 7, \quad 0°, \quad 180°, \quad 0°)$$
$$S(14) = (9.5, \quad 22.5, \quad 2, \quad 0°, \quad 180°, \quad 0°)$$
$$S(15) = (9.5, \quad 22.5, \quad 7, \quad 0°, \quad 180°, \quad 0°)$$

and

$$S(16) = (-7.5, \quad 14.5, \quad 29, \quad 0°, \quad 180°, \quad 0°)$$
$$S(17) = (-7.5, \quad 14.5, \quad 24, \quad 0°, \quad 180°, \quad 0°)$$
$$S(18) = (-7.5, \quad 14.5, \quad 29, \quad 0°, \quad 180°, \quad 0°)$$

**4.8   USE OF VISUAL INFORMATION FOR
           MANIPULATION TASKS**

A robot workstation equipped with a television camera is shown in
Figure 4.15. The aim now is that some of the geometrical parameters of objects
could be attained by processing the signal from a TV camera. Usually only
some of those elements of the transformation matrices which locate the parts
are obtained in this way. Most frequently the position of the centroid (weight
center) and the orientation angles of the chosen coordinate axes on a given
plane are determined from the visual input. The state of a part S_p in the world
coordinate system, the state of camera S_C also in the world coordinate system
and the state of an object S_{PIC} in the coordinate system of the camera as shown
in Figure 4.15 are related by the equation

$$S_p = S_C * S_{PIC} \tag{4.42a}$$

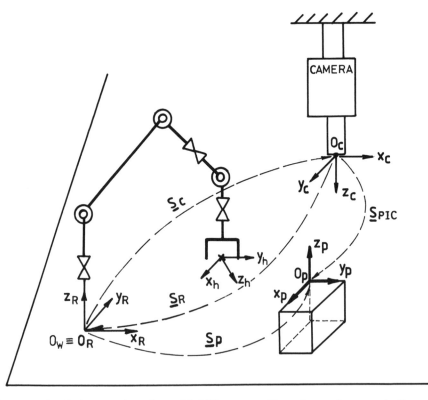

Figure 4.15   Robot workstation with TV camera. Transformations are indicated by
dotted lines.

The transformation matrix S_{PIC} can be inferred from the picture provided by the camera since it is fixed near to the robot in a specified state just outside the boundary of the robot workarea. The matrix S_C is determined for an object being positioned in a known state as

$$S_C = S_P * S_{PIC}^{-1} \tag{4.42b}$$

By knowing the state of the camera from the S_C matrix, we can use it to specify the state of the parts (e.g. a cuboid in Figure 4.15) located at unknown places in the workarea of the robot and given by matrix S_C with respect to the world coordinate system. It is also useful to know that the inverse of S_C equal to S_R is usually used for calibration purposes which represents the transformation from the camera to the origin of the reference coordinate system O_R and expresses how the camera sees the environment in the world coordinate system

$$S_C = S_R^{-1}$$

4.9   *ILLUSTRATIVE EXAMPLE 11*
DESIGN OF A ROBOT WORKSTATION EQUIPPED WITH TV CAMERA

Specification

The picture of the camera contains the origin of the world coordinate system to which the robot is attached. It includes also the reference point of an object to be manipulated by the robot. Let the origin of the reference coordinate system of an assembly part be at O_P and define its relationship to the world coordinate frame by the transformation matrix S_P. The part as seen on the picture will now be represented by the transformation matrix S_{PIC}. The origin of the world system O_R as seen from the camera can be expressed by the transformation matrix S_R.

The coordinates of the reference point of the object (cuboid) as seen from the camera are

$$x = 2 \quad y = 8 \quad z = 5$$

The coordinates of the world origin in arbitrary units as seen from the camera are

$$x = -12 \quad y = 22 \quad z = 8$$

Note that the coordinate frames in the robot workarea are assigned as shown in

Figure 4.15 and the orientation matrix is readily obtainable from the direction of the assigned coordinates, i.e.

$$x_R \| y_P \| x_C \qquad \text{parallel}$$
$$y_R \uparrow\downarrow x_P \uparrow\downarrow y_C \qquad \text{antiparallel}$$
$$z_R \| z_P \uparrow\downarrow z_C \qquad \text{parallel and antiparallel}$$

Design objectives

(i) Develop the transformation matrices S_{PIC} and S_R.
(ii) What is the position of the reference point of the cuboid with respect to the world reference system via the camera position and orientation?
(iii) What is the orientation of the cuboid in the world reference system?

Solutions

(i)

$$S_{PIC} = \begin{bmatrix} 0 & 1 & 0 & 2 \\ 1 & 0 & 0 & 8 \\ 0 & 0 & -1 & 5 \\ 0 & 0 & 0 & 1 \end{bmatrix} \tag{4.43}$$

and

$$S_R = \begin{bmatrix} 1 & 0 & 0 & -12 \\ 0 & -1 & 0 & 22 \\ 0 & 0 & -1 & 8 \\ 0 & 0 & 0 & 1 \end{bmatrix} \tag{4.44}$$

(ii) We use a relationship connecting a series of relative transformations:

$$S(\text{world} \rightarrow \text{cuboid}) = S(\text{world} \rightarrow \text{camera}) * S(\text{camera} \rightarrow \text{cuboid}) \tag{4.45}$$

i.e.

$$S_P = S_R^{-1} * S_{PIC} \tag{4.46}$$

This relationship is readily seen in Figure 4.15. The matrix S_C represents the transformation from the world origin to the camera while S_R is a transformation in the opposite direction.

For computing the inverse of a homogeneous transformation matrix, consider the well-presented mathematical facilities in the literature [4.7] (see Appendix 4.B).

So

$$S_R^{-1} = S_C = \begin{bmatrix} 1 & 0 & 0 & 12 \\ 0 & -1 & 0 & 22 \\ 0 & 0 & -1 & 8 \\ 0 & 0 & 0 & 1 \end{bmatrix} \tag{4.47}$$

and according to Equations 4.42a and 4.46, the S_p becomes

$$S_p = \begin{bmatrix} 1 & 0 & 0 & 12 \\ 0 & -1 & 0 & 22 \\ 0 & 0 & -1 & 8 \\ 0 & 0 & 0 & 1 \end{bmatrix} * \begin{bmatrix} 0 & 1 & 0 & 2 \\ 1 & 0 & 0 & 8 \\ 0 & 0 & -1 & 5 \\ 0 & 0 & 0 & 1 \end{bmatrix}$$

$$= \begin{bmatrix} 0 & 1 & 0 & 14 \\ -1 & 0 & 0 & 14 \\ 0 & 0 & 1 & 3 \\ 0 & 0 & 0 & 1 \end{bmatrix} \tag{4.48}$$

Therefore the position of the reference point of the cuboid is (14, 14, 3) in the world reference system. The x, y and z axes of the coordinate system of the cuboid are parallel to the $-y$, x and z axes of the world reference coordinate system respectively.

4.10   DEVELOPMENT OF THE ROBOT MANIPULATION PROGRAM

In this section we shall demonstrate how the structure of a general robot program can be composed which controls the execution of the tasks outlined in Sections 4.5 and 4.6. For this purpose we introduce an abstract robot language not related to any particular manipulator. We originate transformation expressions in a symbolic code which will then be used to construct a program adaptable, for example, to a robot vision system. We now describe the manipulation with respect to the object in the workspace to be manipulated.

The preparation and explanation of the principal notations are inferred from the robot's environment shown in Figure 4.16.

$$MAN = WP^{-1} * T_6$$

where MAN is the matrix of manipulation, and defines the state of the manipulator with respect to the workpiece. It is determined by the matrices WP and T_6. The purpose of introducing matrix MAN is that we shall be able to imply manipulation with respect to the workpiece at hand and not with respect to the world coordinate system. This greatly simplifies the programming of manipulation and consequently the program length. WP is the matrix describing the position and orientation of the workpiece in the world coordinate system. T_6 is the manipulator transformation matrix and describes the state of the manipulator end point (MEP) in the base coordinate frame.

$$TCP = E$$

where TCP is the tool transformation matrix and describes the state of the tool center point (TCP) with respect to the manipulator end point (MEP). When the gripper is empty, as shown in Figure 4.17, then $TCP = E$. If there is a further tool fixed to the end effector or the gripper holds an object, then $TCP \neq E$ (see Figure 4.18(b)). E is the gripper or tool transformation matrix and describes the size of the gripper/tool as outlined in Section 4.5.

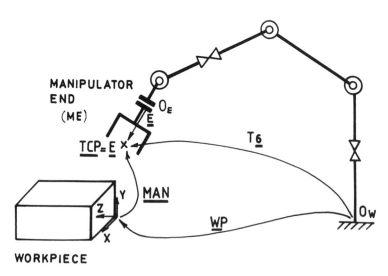

Figure 4.16 Robot's environment for describing the notations of the applied mnemonics.

We now start the program preparation and its explanation in a step-by-step manner and resolve the program development in five steps.

1st step

According to state S(1) the first step of the motion will be started by moving the robot towards the first cuboid denoted by $WP=C[1]$ and shown in Figure 4.17.

$MAN=C[1]^{-1} * T_6$ This defines the state of the manipulator with respect to the first cuboid.

$TCP=E$ While approaching the first cuboid the gripper is empty.

MOVE A_C The gripper moves into the approach position for the first cuboid.

In the further steps of manipulation, i.e. when the second and third cuboid will be moved, there will be no need to change transformation A_C if the cuboids are identical. This can be done only because the manipulation is carried out in the tool coordinate system. The crucial point is that matrix *MAN* is the only thing in the robot program which requires to be changed when the subsequent cuboids will be moved.

2nd step

MOVE G_C *MAN* and *TCP* are unchanged, i.e. the manipulation is still going on with respect to $C[1]$). See S(2) in Figure 4.18(a).

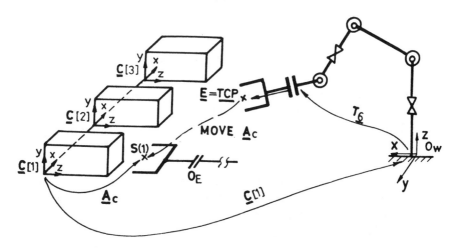

Figure 4.17   The state of the manipulator gripper with respect to the first cuboid's frame.

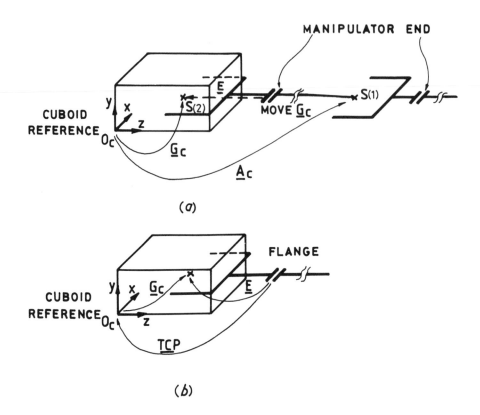

Figure 4.18 States of the gripper. (*a*) Approach position state. (*b*) Grasp position state.

GRASP	After the cuboid has been grasped by the gripper, the *TCP* must be defined.
$TCP = E * G_{\mathrm{C}}^{-1}$	This represents the redefined state of the *TCP* in the cuboid's reference frame (see Figure 4.18(*b*)).

Now the end effector and the workpiece in it are one entity and while describing its further motion we do not have to deal with the workpiece any more. The tool and the workpiece become an integrated part of the manipulator.

3rd step
The target of the third motion step is the frame opening. Recall that the motion is always specified with respect to its target. In order to calculate the new status of *MAN*, we must know the state of the target with respect to the world coordinate frame (see Figure 4.19).

$O[k] = F * O_{\mathrm{F}}[k]$	Matrix $O[k]$ defines the absolute state of the frame opening. $k = 1, 2, 3$ for the three cuboids.

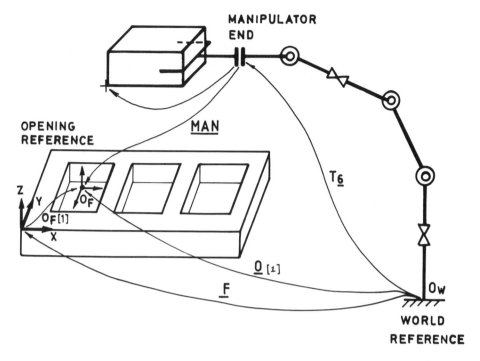

Figure 4.19   States of the frame and its openings.

$MAN = O[k]^{-1} * T_6$ The matrix of the manipulation defines the state of the manipulator end with respect to the opening.

Move CA_0 The robot is approaching the opening (see Figure 4.10(a)).

Move CF_0 The robot inserts the cuboid into the opening.

4th step

Inserting a cuboid into the opening, the frame might have been pushed slightly aside (accommodation). So it may be useful if we determine the new value of matrix F, i.e. the new state of the frame with respect to the world origin. This can be done directly through a TV camera or by reading the actual position of the manipulator joints and recalculating F at the instant of insertion from the following matrix equations.

$O[k] = T_6 * TCP * CF_0^{-1}$ States of the openings derived from the state of the manipulator.

$F = O[k] * O_F[k]^{-1}$ This new value of F will be used when the next cuboid is inserted.

RELEASE   The cuboid is released.

5th step

The cuboid has been released and the gripper must depart from it. The state of the manipulator must now be redefined with respect to the released cuboid. This means that the gripper is empty and $TCP=E$. Thus

$MAN=CF_0^{-1} * O[k]^{-1} * T_6$ See Figure 4.20.
$TCP=E$
MOVE A_C This represents the departure from the cuboid. A_C is the same as in the first step.

The whole program controlling the manipulation of the three cuboids is now listed with short comments:

$TCP=E$ fixing the tool
$k=1$ initializing the loop variable k
(WHILE
$MAN=C[k]^{-1} * T_6$
MOVE A_C
MOVE G_C
GRASP
$TCP=E * G_C^{-1}$ new tool definition

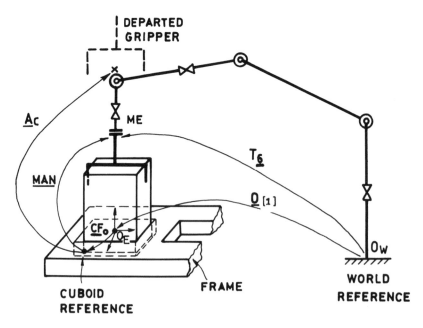

Figure 4.20   The cuboid is in the frame and the gripper in a released state.

$O[k] = F * O_F[k]$	absolute state of the frame opening
$MAN = O[k]^{-1} * T_6$	defining the manipulator with respect to the openings of the frame
MOVE CA_0	approaching the opening
MOVE CF_0	inserting the cuboid
$O[k] = T_6 \times TCP \times CF^{-1}$	
$F = O[k] * O_F[k]^{-1}$	making the state of the frame more accurate by reading $O[k]$ from the image input (the frame may have been displaced during the insertion of the cuboid)
RELEASE	release the cuboid
$MAN = CF_0^{-1} * O[k]^{-1} * T_6$	redefining the manipulator state with respect to the cuboid (see Figures 4.18 and 4.20)
$TCP = E$	TCP relates again to the center point of the gripper
MOVE A_C	departing from the cuboid
$k = k+1$	loop counter is set to the number of the next cuboid
WHILE) $k < 4$	loop executed three times
⋮	
END	end of the program

Comments on the program:

(1) With the appropriate choice of coordinate systems we can minimize the number of those variables which get a new value for every workpiece, i.e. in the above example the $C[k]$ and $O_F[k]$ transformation variables get new values, while the A_C, G_C, CA_0, CF_0 values do not change.

(2) We may include the information provided by the camera by using Equation 4.42a as follows:

READ(PIC)	reads in the state of the cuboid with respect to the camera
$C[k] = S_P * S_{PIC}$	is the absolute state of the cuboid

(3) By redefining the *TCP* we can consider the manipulator and the workpiece held by it as a single item. Thus we can specify the manipulated workpiece itself as argument of motion instructions.

(4) The actual state of the manipulator is defined with respect to the target of the actual motion because the trajectory states (e.g. A_C) along the trajectory leading to the target (e.g. $C[k]$) are defined also with respect to the target.

4.11   GRASP AND TRAJECTORY PLANNING

In Section 4.10 we analyzed a method for creating a robot program consisting of trajectory tracking and pick-and-place operations. The dimensions and locations of objects (workpieces) to be manipulated were supposed to be known either as *a priori* data, for instance, taken from a CAD database, or as data acquired by means of computer vision (see Section 4.9). If data of all dimensions and locations are given before robot programming starts, one would ask how can the sequence of the entire robot motion be automatically generated, or in other words how can the sequence of homogeneous transformation equations and robot motion commands, demonstrated in Section 4.10, be formed under the control of a higher level task description. The different types of information sources needed for generating an automated robot program are described in paragraphs (1) and (2) below.

(1) Types of information source for motion planning

The intelligent abilities of a robot motion planning system depend heavily on the type, amount and possible redundancy of geometrical and technological information available beforehand. Motion planning requires the following types of information:

(a) Model of the robot environment, e.g. the description of the situation in which the robot has to perform its functions. This includes the description of the robot itself.
(b) Model of the workpieces to be manipulated, i.e. the description of objects entering the robotized working process.
(c) Description of the actions, i.e. the operations expected to be performed by the robot on the objects either being in or entering the workspace.

The model of the robot environment, e.g. the description of the situation, comes from the geometrical representation of the working environment, like working cells, assembly table or assembly lines, etc. The exact geometrical description of the free working space of the manipulator can be produced from their representation. The space areas which can be traversed and those prohibited to the robot can be inferred from the situation description. The spatial states of equipment served by the robot, e.g. the states of the machine tools, pallets, fixtures, assembly table, conveyor belt, feeders, etc., can be determined from these sources of information. In addition the situation description provides not only spatial but also time information from the working process, e.g. conveyor speed, cycle time of the served equipment, etc.

These data enable the planner to determine the order of operations performed by the robot and to use all equipment optimally. Information about space and schedule is also of primary importance from the point of view of collision-free trajectory operation [4.8].

The model of the workpieces, i.e. their description, consists of the following factors:

(a) Spatial state (position and orientation) of workpieces: their absolute state or their position and orientation with respect to one another.
(b) Geometry of workpieces: dimensions, geometrical shapes, features, such as mass, moment of inertia, center of gravity, etc.
(c) Further data concerning manipulation: grasp points, base or support faces etc.

The more automatic we require the motion of the manipulator to be, the more detailed must be the description of the workpiece. Exact knowledge about workpiece geometry is necessary if the motion sequence to grasp the workpiece by the end effector has to be done with no human interaction or guidance. The knowledge of the relative position and orientation of workpieces makes it possible to automatically infer the order and the total motion requirement for a robot assembly process [4.1].

The description of the actions consists of an explicit assembly plan which is often needed since the formation of such a plan can be carried out automatically only in special cases. In the framework of operation description we have to create a plan which matches the level of intelligence (inferring ability) of the available assembly planning system. The more detailed the description of the sequence of robot actions by means of a programming language, the lower will be the number of logical inference decisions to be carried out by the motion planning system. On the basis of the actual situation, workpiece and operation descriptions, the task of the trajectory planning system is to complete the series of robot motions needed to accomplish the given manipulation task and other auxiliary operations in connection with the manipulation, e.g. the operation of tools, grippers, fixtures, feeders, etc.

The process of motion planning can be compared to the translation of high level computer programming languages by a compiler. The high level motion program (compiled by the planner) consists of symbolic data references, while the target program is in terms of the physical features of the robot arm and refers to its functional operation. The resulting robot program, which runs in an interpretative mode, is similar to the machine code language of computers.

(2) Geometrical database

Motion planning systems usually consist of a geometrical modeler or some commonly used modeling package embedded in a high level robot programming language which the user can make use of. The geometrical modeler enables the programmer to visualize and include the environment of the robot, the parts for the assembly to be put together and the robot itself. There exist a number of modelers; some of them are commercially available. In the following, we present, as an example, a typical geometrical construction package called MODBUILD (MODel BUILDer) [4.9]. This, and surface-oriented geometric modeler, have the following three main components:

(a) Viewing: the wireframe model of the body under construction can be seen on display from arbitrary viewpoints, from any distance (perspective views) and in any scale. These parameters are set by adjusting a virtual camera.

(b) Hidden line removal: in any phase of the body construction and in any camera setting, the model can be displayed with hidden lines removed. Hidden lines can be also correctly eliminated if a number of bodies cover each other. The hidden line algorithm works also on curved surfaces such as spheres, cylinders and cones.

(c) Body construction: the computer representation of bodies is in three dimensions accomplished according to a vertex–edge–surface hierarchy. When constructing complex bodies one has the possibility of building them from simpler bodies by merging body descriptions.

MODBUILD is a surface modeler and not a solid modeler like, for example, BUILD. Any kind of linear transformation can be performed on models by using 4×4 homogeneous transformation matrices. In the case of primary modeling one starts with defining the vertices of the body. Vertices can be connected then by straight or circular edges. The sequence of edges will determine the surfaces. Smoothly joining surfaces are connected by fictive edges which are not shown on hidden-line pictures.

The use of MODBUILD is easy to learn. It is controlled by means of a hierarchical control tree. Command menus help the operator to select the appropriate operation. The user can point at the desired geometrical feature by using the cursor of a graphics display.

MODBUILD enables the user to input specific information concerning the manipulation of the body in question. Such supplementary information can be assigned to the body as a whole and to any of its surfaces individually, e.g. grasp points, approach directions, support surfaces, grasping force, etc.

As an example of a complex body, Figure 4.21 shows the model of a robot gripper in which one can see the wireframe representation of the gripper.

Figure 4.22 shows a complex spatial situation described by MODBUILD. The cylindrical body, the fixture and the manipulator constituting the assembled workpiece are independently modeled. Their relative positions and orientations are set by a higher level program called ANIMATOR [4.9] which controls the motion of bodies by setting the transformation features of MODBUILD.[1]

ANIMATOR is not only a robot simulator but also a teaching aid for robot programming. Individual joints as well as the end effector of the robot model can be guided interactively or under control by specific robot language instructions. These types of instructions represent simple motion sequences, e.g. the motion of the robot gripper along a specified straight-line section expressed in cartesian coordinates or the rotation of the gripper about one of the selected axes. These compact instructions are then decomposed into more elementary motion control commands by robot dependent and application oriented routines. Parameters of previously defined motion programs can be modified. The robot motion simulation package can demonstrate the execution of robot programs at different step resolutions as well as the starting of a motion from different initial gripper states or robot configurations.

The virtual camera parameters can also be modified and enable the user to check the same motion phases from different viewing directions. A motion sequence of an articulated six-jointed manipulator, the Corohand robot, developed by the above packages, is shown in Figure 4.23. The motion sequence of this robot arm was generated by a higher level program that calculates trajectories [4.10], [4.11]. The use of this kind of graphic motion simulator reduces the risk caused by newly created robot programs, since wrong motion commands causing collisions can be detected before the program is put on the robot itself.

Note that an affix mechanism [4.5] is also incorporated into the system. The operator specifies which object from a set has to be moved by the robot and the system automatically sets its appropriate spatial coordinates.

4.12 APPENDICES

Appendix 4.A Derivation of the states of the manipulator end S(2), S(3) and S(4)

The resultant matrices $T_6(1)$, $T_6(2)$, $T_6(3)$ and $T_6(4)$ are already given in Equations 4.41.

---

[1]MODBUILD and ANIMATOR have been designed at the Computer and Automation Institute of the Hungarian Academy of Sciences.

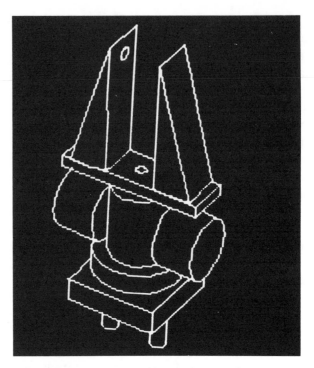

Figure 4.21 Model of robot end effector.

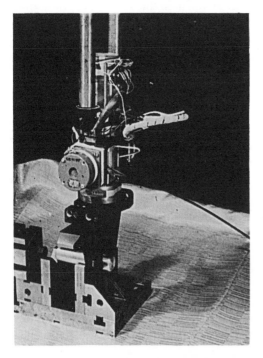

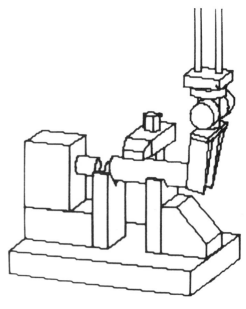

Figure 4.22 Spatial situation modeling.

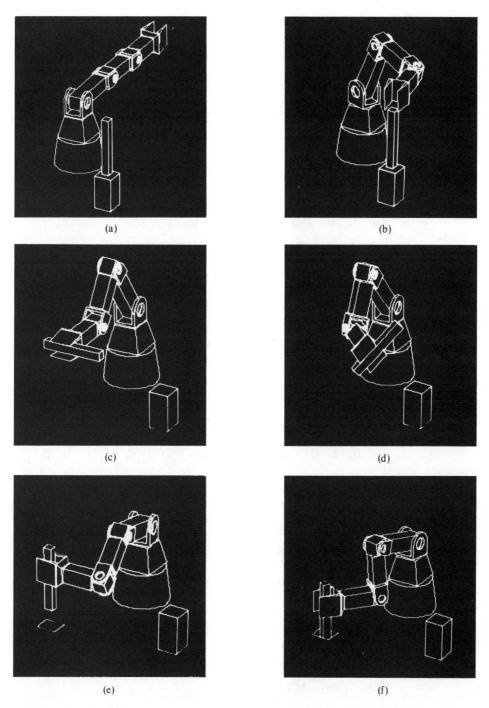

Figure 4.23 Demonstration of a CAD motion sequence of an articulated six-jointed manipulator. (*a*) Initial position. (*b*) Before gripping. (*c*) In motion. (*d*) Above target. (*e*) Final orientation. (*f*) Final position.

The proof of the latter three is given below:

$$T_6(2) = C[2] * G_C * E^{-1}$$

$$= \begin{bmatrix} 0 & 0 & -1 & 500 \\ -1 & 0 & 0 & 300 \\ 0 & 1 & 0 & 0 \\ 0 & 0 & 0 & 0 \end{bmatrix} * \begin{bmatrix} -1 & 0 & 0 & 20 \\ 0 & 1 & 0 & 20 \\ 0 & 0 & -1 & 110 \\ 0 & 0 & 0 & 1 \end{bmatrix} * \begin{bmatrix} 1 & 0 & 0 & 0 \\ 0 & 1 & 0 & 0 \\ 0 & 0 & 1 & -160 \\ 0 & 0 & 0 & 1 \end{bmatrix}$$

$$= \begin{bmatrix} 0 & 0 & 1 & 230 \\ 1 & 0 & 0 & 280 \\ 0 & 1 & 0 & 20 \\ 0 & 0 & 0 & 1 \end{bmatrix}$$

$$T_6(3) = F * O_F[1] * CA_0 * G_C * E^{-1}$$

$$= \begin{bmatrix} -1 & 0 & 0 & 100 \\ 0 & -1 & 0 & 400 \\ 0 & 0 & 1 & 10 \\ 0 & 0 & 0 & 1 \end{bmatrix} * \begin{bmatrix} 0 & 1 & 0 & 31 \\ -1 & 0 & 0 & 31 \\ 0 & 0 & 1 & 0 \\ 0 & 0 & 0 & 1 \end{bmatrix}$$

$$* \begin{bmatrix} 0 & 1 & 0 & 20 \\ -1 & 0 & 0 & 20 \\ 0 & 0 & 1 & 100 \\ 0 & 0 & 0 & 1 \end{bmatrix} * \begin{bmatrix} -1 & 0 & 0 & 20 \\ 0 & 1 & 0 & 20 \\ 0 & 0 & -1 & 110 \\ 0 & 0 & 0 & 1 \end{bmatrix} * \begin{bmatrix} 1 & 0 & 0 & 0 \\ 0 & 1 & 0 & 0 \\ 0 & 0 & 1 & -160 \\ 0 & 0 & 0 & 1 \end{bmatrix}$$

$$= \begin{bmatrix} 0 & -1 & 0 & 69 \\ 1 & 0 & 0 & 369 \\ 0 & 0 & 1 & 10 \\ 0 & 0 & 0 & 1 \end{bmatrix} * CA_0 * G_C * E^{-1}$$

$$= \begin{bmatrix} 1 & 0 & 0 & 49 \\ 0 & 1 & 0 & 389 \\ 0 & 0 & 1 & 100 \\ 0 & 0 & 0 & 1 \end{bmatrix} * G_C * E^{-1}$$

$$= \begin{bmatrix} 1 & 0 & 0 & 69 \\ 0 & 1 & 0 & 409 \\ 0 & 0 & -1 & 220 \\ 0 & 0 & 0 & 1 \end{bmatrix} * E^{-1}$$

$$= \begin{bmatrix} -1 & 0 & 0 & 69 \\ 0 & 1 & 0 & 409 \\ 0 & 0 & -1 & 380 \\ 0 & 0 & 0 & 1 \end{bmatrix}$$

$$T_6(4) = F * O_F[1] * CF_0 * G_C * E^{-1}$$

$$= \begin{bmatrix} 0 & -1 & 0 & 69 \\ 1 & 0 & 0 & 369 \\ 0 & 0 & 1 & 10 \\ 0 & 0 & 0 & 1 \end{bmatrix} * \begin{bmatrix} 0 & 1 & 0 & 20 \\ -1 & 0 & 0 & 20 \\ 0 & 0 & 1 & -10 \\ 0 & 0 & 0 & 1 \end{bmatrix}$$

$$F * O_F[1]$$

$$* \begin{bmatrix} -1 & 0 & 0 & 20 \\ 0 & 1 & 0 & 20 \\ 0 & 0 & -1 & 110 \\ 0 & 0 & 0 & 1 \end{bmatrix} * \begin{bmatrix} 1 & 0 & 0 & 0 \\ 0 & 1 & 0 & 0 \\ 0 & 0 & 1 & -160 \\ 0 & 0 & 0 & 1 \end{bmatrix}$$

$$= \begin{bmatrix} 1 & 0 & 0 & 49 \\ 0 & 1 & 0 & 389 \\ 0 & 0 & 1 & 0 \\ 0 & 1 & 0 & 1 \end{bmatrix} G_C * E^{-1}$$

$$= \begin{bmatrix} -1 & 0 & 0 & 69 \\ 0 & 1 & 0 & 409 \\ 0 & 0 & -1 & 110 \\ 0 & 0 & 0 & 1 \end{bmatrix} * E^{-1}$$

$$T_6(4) = \begin{bmatrix} -1 & 0 & 0 & 69 \\ 0 & 1 & 0 & 409 \\ 0 & 0 & -1 & 270 \\ 0 & 0 & 0 & 1 \end{bmatrix}$$

Appendix 4.B

According to Equation 2.5 the general form of a transformation matrix is

$$H = \left[\begin{array}{ccc|c} e_1 & e_2 & e_3 & k \\ \hline 0 & 0 & 0 & 1 \end{array} \right] = \left[\begin{array}{c|c} R & k \\ \hline 0 \quad 0 \quad 0 & 1 \end{array} \right] \tag{4.B.1}$$

where

$$e_i = [e_{ix} \; e_{iy} \; e_{iz}]^T$$

and

$$i = 1, 2, 3$$

and the inversion of matrix H results

$$H^{-1} = \begin{bmatrix} R & k \\ 0^T & 1 \end{bmatrix}^{-1} = \left[\begin{array}{c|c} R^{-1} & -R^{-1}k \\ \hline 0^T & 1 \end{array} \right]$$

In robot coordinate transformation there is a great demand for computing inverse matrices for both direct and inverse transformations [4.12]. It is known that the inverse of the rotation submatrix R^{-1} equals its transpose but the inverse of a homogeneous transformation submatrix is not equal to its transpose. A short and powerful method for matrix inversion is shown as

$$H^{-1} = \begin{bmatrix} e_{1x} & e_{1y} & e_{1z} & -e_1^T k \\ e_{2x} & e_{2y} & e_{2z} & -e_2^T k \\ e_{3x} & e_{3y} & e_{3z} & -e_3^T k \\ 0 & 0 & 0 & 1 \end{bmatrix} \tag{4.B.2}$$

where $e_1^T k$ represents the scalar product of the vectors e_1 and k. Then

$$e_1^T k = [e_{1x}\ e_{1y}\ e_{1z}] \begin{bmatrix} k_x \\ k_y \\ k_z \end{bmatrix} = e_{1x}k_x + e_{1y}k_y + e_{1z}k_z$$

$$e_2^T k = [e_{2x}\ e_{2y}\ e_{2z}] \begin{bmatrix} k_x \\ k_y \\ k_z \end{bmatrix} = e_{2x}k_x + e_{2y}k_y + e_{2z}k_z$$

$$e_3^T k = [e_{3x}\ e_{3y}\ e_{3z}] \begin{bmatrix} k_x \\ k_y \\ k_z \end{bmatrix} = e_{3x}k_x + e_{3y}k_y + e_{3z}k_z \tag{4.B.3}$$

To demonstrate a homogeneous matrix inversion by a numerical example find the inverse of the following transformation matrix H and prove it is its inverse.

$$H = \begin{bmatrix} 1 & 0 & 0 & 2 \\ 0 & 1 & 0 & 3 \\ 0 & 0 & -1 & 4 \\ 0 & 0 & 0 & 1 \end{bmatrix} \tag{4.B.4}$$

Applying Equation 4.B.2 we readily obtain

$$H^{-1} = \begin{bmatrix} 1 & 0 & 0 & -2 \\ 0 & 1 & 0 & -3 \\ 0 & 0 & -1 & 4 \\ 0 & 0 & 0 & 1 \end{bmatrix} \tag{4.B.5}$$

If this solution is correct, we evidently require that $HH^{-1} = I$

$$HH^{-1} = \begin{bmatrix} 1 & 0 & 0 & 2 \\ 0 & 1 & 0 & 3 \\ 0 & 0 & -1 & 4 \\ 0 & 0 & 0 & 1 \end{bmatrix} * \begin{bmatrix} 1 & 0 & 0 & -2 \\ 0 & 1 & 0 & -3 \\ 0 & 0 & -1 & 4 \\ 0 & 0 & 0 & 1 \end{bmatrix}$$

$$= \begin{bmatrix} 1 & 0 & 0 & 0 \\ 0 & 1 & 0 & 0 \\ 0 & 0 & 1 & 0 \\ 0 & 0 & 0 & 1 \end{bmatrix} = I \qquad (4.B.6)$$

We note that this matrix inversion method is applicable only for homogeneous matrices but not valid for arbitrary 4 by 4 matrices.

4.13 REFERENCES

[4.1] Ambler, A. P. and Poppleston, R. J. Inferring of position of bodies from specified spatial relationship, *Artificial Intelligence*, Vol. 6, 1975.

[4.2] Popplestone, R. J., Ambler, A. B. and Bellos, I. RAPT–A language for describing assemblies, *The Industrial Robot*, Vol. 5, No. 3, Sept. 1978, pp. 113–37.

[4.3] Lozano-Perez, T. Automatic planning of manipulator transfer movements, *IEEE Trans. on Systems, Man and Cybernetics*, Vol. SMC-11, No. 10, 1981.

[4.4] Taylor, R. H. The synthesis of manipulator control programs from task-level specifications, PhD Thesis, Stanford University, 1976.

[4.5] Users Guide to VAL: A robot programming and control system, *Unimation Robotics*, June 1980.

[4.6] Nevins, J. L. and Whitney, D. E. Computer controlled assembly, *Scientific American*, Feb. 1978.

[4.7] Lee, C. S. G. Robot arm kinematics and control, *Computer*, Dec. 1982.

[4.8] Udupa, S. M. Collision detection and avoidance in computer-controlled manipulators, PhD Thesis, California Institute of Technology, Sept. 1976.

[4.9] Bathor, M. and Siegler, A. Graphic simulation for robot programming, *Proc. of the 4th British Robot Association Annual Conference*, Brighton, UK, 1981, pp. 85–93.

[4.10] Siegler, A. and Zilahy, F. Computer control of a six degrees of freedom articulated manipulator, *Proc. of MANUFACONT*, *80*, Budapest, Hungary, 1980, pp. 161–7.

[4.11] Siegler, A. Kinematics and microcomputer control of a six degrees of freedom manipulator, *Research Report*, Cambridge University Engineering Department, CUED-CMS 185/1979.

[4.12] Snyder, W. E. *Industrial Robots: Computer Interfacing and Control*, Prentice-Hall, 1985.

5

MODELING OF MANIPULATOR GEOMETRY

5.1   REPRESENTATION OF MANIPULATOR GEOMETRY

In this chapter we will analyze various robot manipulator constructions and formulate methods for assigning coordinate systems to serial linkages used to model the serial robot construction. In Chapters 2 and 3 we derived formulas for defining position and orientation of the coordinate frames in a cartesian workspace. We will now apply these formulas to the manipulator's end effector and express its states in terms of the joint variables of the manipulator itself.

As outlined in Section 2.6, the serial robot manipulator consists of a series of links activated by joints. To describe the relationship between the links we will use the homogeneous transformation to express the states of the end effector in terms of manipulator joint variables. Recalling Equations 2.15 and 2.16 the state of the ith link with respect to the base was given as

$$H_{0,i} = H_{0,1} * \ldots * H_{i-1,i} \tag{5.1}$$

where $H_{0,1}$ describes the state of the first link with respect to the base, ..., $H_{i-1,i}$ describes the state of the last link with respect to the previous one.

5.2   SPECIFYING THE HAND GRIPPER STATES

The function of a manipulator is to carry the hand or the tool with arbitrary orientation to any point within the robot workspace. To fulfil this task it must have at least six controllable joints, the controlled configuration of which can provide the desired six end effector variables, three cartesian and three orientation parameters. In Figure 5.1 x_h, y_h, z_h are the cartesian coordinates of the gripper or tool origin O_h with respect to the base. The Euler angles α_h, β_h, γ_h in Equation 5.2 represent the orientation of the hand as described in Section 3.1. So the state of the hand S_h in implicit form with respect to the base yields

$$S_h = f(x_h, y_h, z_h, \alpha_h, \beta_h, \gamma_h) \tag{5.2}$$

This sextuple consists of six independent variables, which will fully determine the position and orientation of the unit vector triad of the hand coordinate system (e_1, e_2, e_3) with respect to the base coordinate origin O_0 shown in Figure 5.1.

To determine the end effector's arbitrary state we have to provide a specified number of independent variables in the hand (tool) coordinate system. In order to control the hand by its variables, we need to have at least as many

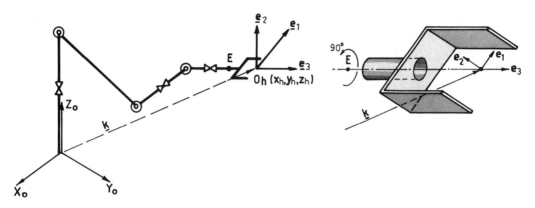

Figure 5.1 State of the hand gripper in the base coordinate frame.

controllable robot variables as we need to control the end effector (gripper or tool).

To bring the manipulator end effector into a given state characterized by the relationship in Equation 5.2, we need a minimum six degrees of freedom manipulator, i.e. six independently controllable joints of the appropriate arrangement, which can be positioned anywhere within their own motion limits. These can be obtained, for instance, by a serial six-jointed manipulator whose state is expressed by the overall transformation matrix T_6, namely

$$T_6 = H_{0,1} * H_{1,2} * H_{2,3} * H_{3,4} * H_{4,5} * H_{5,6} \tag{5.3}$$

Note that a controllable joint does not necessarily represent a degree of freedom in itself, but for position and/or orientation controllability it must be an independent joint (see Section 1.8).

The number of controllable end effector variables for a given manipulator also depends on the kinematic configuration of the joints. If, for instance, only the position of the end effector is of significance, a minimum of three controllable joints are needed and in addition, to determine the orientation of the end effector, another three controllable joints would be necessary.

The state of the manipulator is expressed in terms of the position and orientation of its end effector. The origin of the end effector's coordinate system is located centrally between the fingertips at tool center point (TCP) and the vector k represents the position of this center point with respect to the origin O_0 (see Figure 5.1).

The orientation of the hand is described by the three unit orthogonal vectors (e_1, e_2, e_3) of a given coordinate system along the x, y, z axes respectively. They are defined as follows. The $e_3 (z$ axis) points in the direction in which the hand approaches the object to be grasped and therefore it may be called the approach vector. The $e_1 (x$ axis) points from one fingertip to the other and therefore it is usually called the orientation vector. Finally $e_2(y$ axis) is specified by the cross-product in Equation 5.4 and therefore becomes the normal vector since it is perpendicular to the plane of the fingers.

$$e_2 = e_3 \times e_1 \tag{5.4}$$

The elements of the manipulator transformation matrix T_6 in Equation 5.3 can be computed as outlined in Section 4.6.

$$T_6 = \begin{bmatrix} e_{1x} & e_{2x} & e_{3x} & k_x \\ e_{1y} & e_{2y} & e_{3y} & k_y \\ e_{1z} & e_{2z} & e_{3z} & k_z \\ 0 & 0 & 0 & 1 \end{bmatrix} \tag{5.5}$$

On the other hand, the orientation of the hand gripper can be expressed in terms of the Euler angles. Euler angles for roll–yaw–roll, roll–pitch–yaw and roll–pitch–roll geometries were discussed in Section 3.1. Now it will be shown how the Euler angles are implemented, for instance for the roll–yaw–roll configuration with reference to the manipulator's end effector (see Figure 5.2).

Now the aim is to express the hand's reference coordinates, i.e. the tool coordinates. This process is carried out in three steps as shown in Figure 5.2.

(i) A rotation by angle $-\gamma_h$ is needed so that the hand's own plane (ABO_h) becomes parallel with the z_0 axis of the base coordinate system (rotation about z_h axis).

(ii) A rotation $-\beta_h$ is needed to make the z_h approach axis of the end effector parallel with the z_0 axis of the base coordinate system (rotation about x_0 axis).

(iii) Finally, a rotation $-\alpha_h$ is needed to make the orientation vector pointing from fingertip to fingertip, denoted by x'_h, parallel with the x_0 axis of the base coordinate system (rotation about z_0 axis).

The rotation sequence shown graphically in Figure 5.2, demonstrates clearly that these rotations are not commutative. This can easily be proved by checking the product of their corresponding orientation matrices when multiplying them together in a different order.

For instance if we choose the sequence $(\alpha_h, \beta_h, \gamma_h)$ and by introducing the abbreviations $\cos \alpha_h = c\alpha_h$ and $\sin \alpha_h = s\alpha_h$, etc., then the resultant matrix is

$$R(\alpha_h, \beta_h, \gamma_h) = \begin{bmatrix} c\alpha_h c\gamma_h - s\alpha_h c\beta_h s\gamma_h & -c\alpha_h s\gamma_h - s\alpha_h c\beta_h c\gamma_h & s\alpha_h s\beta_h \\ s\alpha_h c\gamma_h + c\alpha_h c\beta_h s\gamma_h & -s\alpha_h s\gamma_h + c\alpha_h c\beta_h c\gamma_h & -c\alpha_h s\beta_h \\ s\beta_h s\gamma_h & s\beta_h c\gamma_h & c\beta_h \end{bmatrix} \tag{5.6}$$

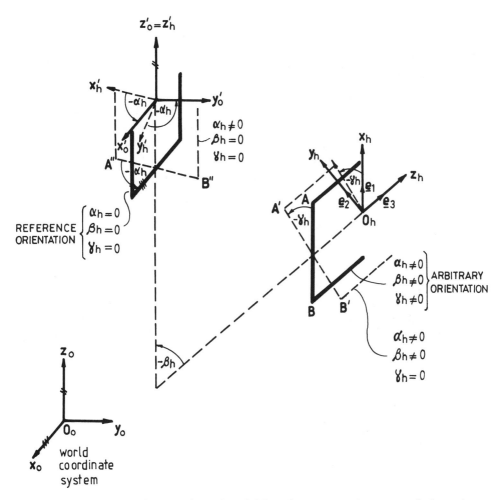

Figure 5.2 Manipulator orientation with reference to the state of the gripper (roll–yaw–roll).

However, if we change over the angles β_h and γ_h, the sequence of the rotations will be $(\alpha_h, \gamma_h, \beta_h)$ and the resultant transformation matrix is

$$R_{RYR}(\alpha_h, \gamma_h, \beta_h) = \begin{bmatrix} \cos(\alpha_h + \gamma_h) & -\cos\beta_h \sin(\alpha_h + \gamma_h) & \sin\beta_h \sin(\alpha_h + \gamma_h) \\ \sin(\alpha_h + \gamma_h) & \cos\beta_h \cos(\alpha_h + \gamma_h) & -\sin\beta_h \cos(\alpha_h + \gamma_h) \\ 0 & \sin\beta_h & \cos\beta_h \end{bmatrix}$$

$$(5.7)$$

The resultant hand orientation matrices in Equations 5.6 and 5.7 make it quite clear why these rotations are not commutative.

5.3   MODELING OF A SYMMETRICALLY STRUCTURED ROBOT BY *H* MATRICES

We will now consider how to develop the transformation matrices in Equation 5.1 for a serial link robot manipulator, which consists of six joints. The only significance of a link is that it keeps a fixed relationship between the joints at both ends. The characteristic link and joint parameters were discussed in Section 2.4 where we developed the tranformation matrices of revolute and prismatic joints (see Figure 2.10). Here we will discuss how these simple transformation matrices are brought together to provide the resultant transformation matrix for the overall robot manipulation.

The first step in numerical modeling of a robot manipulator is to classify its types of joints. For this purpose we will choose the vertically symmetrical six-jointed Hitachi assembly robot shown in Figure 5.3.

The Hitachi assembly robot is a symmetrically structured revolute type of mechanism, i.e. it is an RRRRRR or briefly 6R mechanism. Having determined the type of mechanism, the next step is to specify the link parameters (also see Figure 5.5).

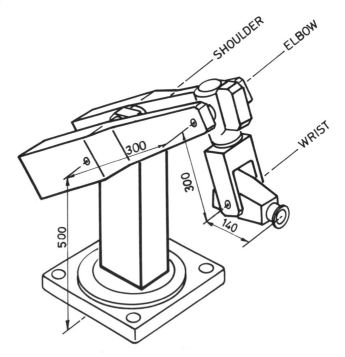

Figure 5.3   Vertically symmetrical Hitachi assembly robot.

The length of each link is denoted as follows:

LINK L0 (base–shoulder)   A
LINK L1 (shoulder)   0
LINK L2 (shoulder–elbow) B
LINK L3 (elbow–wrist)   C
LINK L4 (wrist)   0
LINK L5 (wrist)   0
LINK L6 (wrist–end)   D

The next step is to describe the motion of the joints. Every joint motion can be described by a single scalar variable. We will distinguish between the two types of revolute joints:

(1) Swivel or twisting joint (rotation about link axis).
(2) Bend or tilting joint (rotation about axis transverse to link axis).

There are many revolute robots which may have prismatic joints built in, usually between the elbow and the first wrist joint. The symbols of all three joints are shown in Figure 5.4.

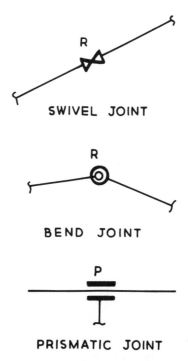

Figure 5.4   Symbols of various types of joints.

Both swivel and bend joints are one degree of freedom joints [Fig. 5.4]. The linkage theory does not distinguish between the various types of revolute joints [5.1]. However, we will make a distinction so that the various constant parameters and the controlled joint variables required in the derivation of their transformation matrices could be separately available as tabulated in Tables 5.1 and 5.2.

Usually the robot has a reference (home) position. For the description of a robot let us first consider the Hitachi assembly robot in its reference position as shown in Figure 5.5. The joint variables are designated q_i. In this configuration each link is perpendicular to the base, which means that all links are aligned in the vertical direction and the manipulator points upwards. Thus the reference position of the TCP for (x, y, x) is

$$(0, 0, A + B + C + D)$$

Table 5.1 Parameters and variables of the Hitachi assembly robot.

	Link number						
Variable	*L0*	*L1*	*L2*	*L3*	*L4*	*L5*	*L6*
α_i^*	–	$0°$	q_2	q_3	$0°$	q_5	$0°$
ϑ_i†	–	q_1	$0°$	$0°$	q_4	$0°$	q_6
a_i	0	0	0	0	0	0	0
b_i	A	0	B	C	0	0	D

*$\alpha_i = [q_i]$ revolute or transversal motion
†$\vartheta_i = [q_i]$ swivel or longitudinal motion

Table 5.2 Parameters and variables of the Unimation PUMA robot.

	Link number						
Variable	*L0*	*L1*	*L2*	*L3*	*L4*	*L5*	*L6*
α_i^*	–	$0°$	q_2	q_3	$0°$	q_5	$0°$
ϑ_i†	–	q_1	$0°$	$0°$	q_4	$0°$	q_6
a_i	0	S	T	0	0	0	0
b_i	0	0	B	C	0	0	0

*$\alpha_i = [q_i]$ = revolute or transversal motion
†$\vartheta_i = [q_1]$ = swivel or longitudinal motion

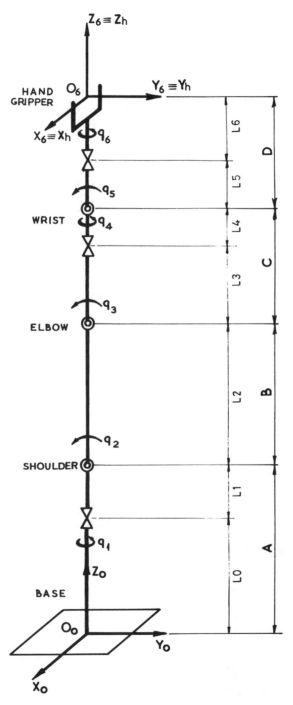

Figure 5.5 Reference state of a revolute robot.

and the reference orientation of the gripper in terms of Euler angles $(\alpha_h, \beta_h, \gamma_h)$ is

$$(0°, 0°, 0°)$$

Let us now assign the so-called task-oriented commands to the axes of the hand coordinate system. In the reference configuration the SWEEP motion is parallel to the x_0 axis, the REACH motion parallel to the z_0 axis and the LIFT motion is perpendicular to the axes x_0 and z_0, i.e. in the direction of y_0 (see Figure 5.6).

The coordinate frames assigned to the joints of the Hitachi assembly robot in an arbitrary position are shown in Figure 5.7. When we compare the assignment of these coordinate systems to that in Figure 2.11, it becomes clear that the a_i values in the matrix in Equation 2.29 are zero for every i since the z_i axes are always aligning with the ith link. This symmetrical property of the manipulator geometry at hand will be convenient, since the upper two elements of the right column in the joint transformation matrix in Equation 2.29 will vanish.

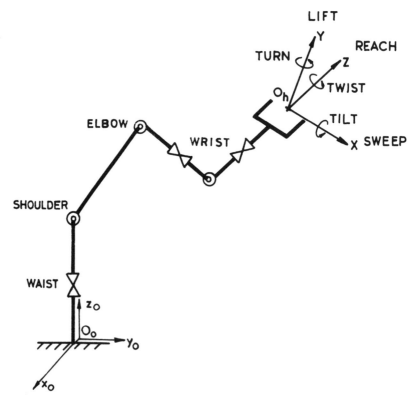

Figure 5.6   Task-oriented gripper commands.

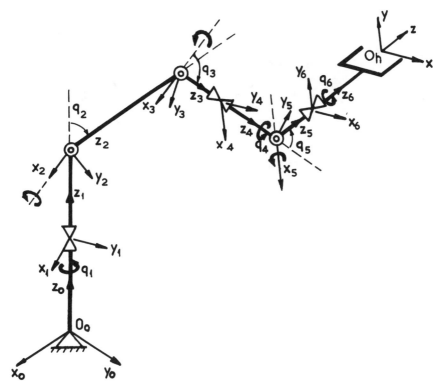

Figure 5.7 Coordinate frames assigned to the joints of the Hitachi assembly robot in Figure 5.3.

According to the notation in Figure 5.5, the rotations q_1, q_4 and q_6 turn the corresponding links in a counterclockwise direction about the z_1, z_4 and z_6 axes respectively, i.e. in home position about the z_0 axis. The rotations q_2, q_3 and q_5, when they occur in the positive direction, bend the corresponding joints about the x_2, x_3 and x_5 axes also in a counterclockwise direction respectively, i.e. in home position about the x_0 axis. The rotations of q_1, q_4 and q_6 perform swivel motions about their longitudinal axes (z) and q_2, q_3 and q_5 perform bending motions about their transverse axes (x). In the first case the axis of rotation aligns with the joining links while in the second case it is perpendicular (transverse) to them [5.2] as shown in Figure 5.8.

The transformation matrix of the swivel joint rotation about its longitudinal axis z_j at a distance b_j from the axis of rotation is given by

$$
H_{j-1,j} = \begin{bmatrix} \cos q_j & -\sin q_j & 0 & 0 \\ \sin q_j & \cos q_j & 0 & 0 \\ 0 & 0 & 1 & b_j \\ 0 & 0 & 0 & 1 \end{bmatrix} \tag{5.8}
$$

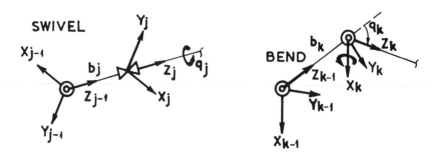

Figure 5.8   Single joint-link configurations.

The transformation matrix of the bend joint rotation about its transverse axis x_k, where the link length is b_k, is given by

$$
H_{k-1,k}=
\begin{bmatrix}
1 & 0 & 0 & 0 \\
0 & \cos q_k & -\sin q_k & 0 \\
0 & \sin q_k & \cos q_k & b_k \\
0 & 0 & 0 & 1
\end{bmatrix}
\tag{5.9}
$$

Note that in a simple robot geometry the upper two elements of the last columns in Equations 5.8 and 5.9 are zero, because there are no offsets. The actual position of the origin of a swivel joint along the link connecting two bending joints has no significance.

Now, when advancing from the end of the manipulator to the base, the following transformations are performed (see Figure 5.5):

(1) A shift in the z direction and a rotation about z D, q_6
(2) A rotation about x q_5
(3) A shift in the z direction and a rotation about z C, q_4
(4) A rotation about x q_3
(5) A shift in the z direction B
(6) A rotation about x q_2
(7) A shift in the z direction and a rotation about z A, q_1

The constant parameters and the joint variables of the Hitachi assembly manipulator are listed in Table 5.1. Note that this robot has no constant parameters because of its symmetry, but in robot computation sometimes the 'no or nothing' could also be significant.

By substituting the motion variables of the individual joints in the matrix into Equation 2.29 we get the following transformations:

$$H_{0,1} = \begin{bmatrix} \cos q_1 & -\sin q_1 & 0 & 0 \\ \sin q_1 & \cos q_1 & 0 & 0 \\ 0 & 0 & 1 & A \\ 0 & 0 & 0 & 1 \end{bmatrix} \tag{5.10}$$

$$H_{1,2} = \begin{bmatrix} 1 & 0 & 0 & 0 \\ 0 & \cos q_2 & -\sin q_2 & 0 \\ 0 & \sin q_2 & \cos q_2 & 0 \\ 0 & 0 & 0 & 1 \end{bmatrix} \tag{5.11}$$

$$H_{2,3} = \begin{bmatrix} 1 & 0 & 0 & 0 \\ 0 & \cos q_3 & -\sin q_3 & 0 \\ 0 & \sin q_3 & \cos q_3 & B \\ 0 & 0 & 0 & 1 \end{bmatrix} \tag{5.12}$$

$$H_{3,4} = \begin{bmatrix} \cos q_4 & -\sin q_4 & 0 & 0 \\ \sin q_4 & \cos q_4 & 0 & 0 \\ 0 & 0 & 1 & C \\ 0 & 0 & 0 & 1 \end{bmatrix} \tag{5.13}$$

$$H_{4,5} = \begin{bmatrix} 1 & 0 & 0 & 0 \\ 0 & \cos q_5 & -\sin q_5 & 0 \\ 0 & \sin q_5 & \cos q_5 & 0 \\ 0 & 0 & 0 & 1 \end{bmatrix} \tag{5.14}$$

$$H_{5,6} = \begin{bmatrix} \cos q_6 & -\sin q_6 & 0 & 0 \\ \sin q_6 & \cos q_6 & 0 & 0 \\ 0 & 0 & 1 & D \\ 0 & 0 & 0 & 1 \end{bmatrix} \tag{5.15}$$

In Equation 5.15 D represents the distance from joint J6 to the TCP. If the end effector is changeable, D can be the distance from joint J6 to the manipulator end (wrist flange).

5.4   MODELING OF AN ASYMMETRICALLY STRUCTURED ROBOT BY *H* MATRICES

We will now consider another robot with somewhat different joint configuration and develop the transformation of its joints. For this purpose we will choose the Unimation PUMA robot shown in Figure 5.9 and its joint coordinates shown in Figure 5.10. It can be seen that it has an asymmetric kinematic geometry, i.e. there are two offsets, one at the intersection of the shoulder's transverse axis and the waist's longitudinal axis designated by S and another at the elbow denoted by T.

The constant parameters and the joint variables of the Unimation PUMA manipulator to be used in the transformations are listed in Table 5.2. (Note that there are the two constant offsets S and T.) A coordinate frame assigned to each joint of the PUMA robot is shown in Figure 5.10.

If we place the origin of the reference system on the transverse axis of the shoulder, then the length of the link L0 is zero ($A=0$) as listed in Table 5.2.

In the case of the PUMA, the origin of the last three joint axes is in the wrist and the manipulator end can be placed at this origin as well. Thus the lengths of the wrist links are zero ($D=0$). In so doing we get the following transformations at the joints

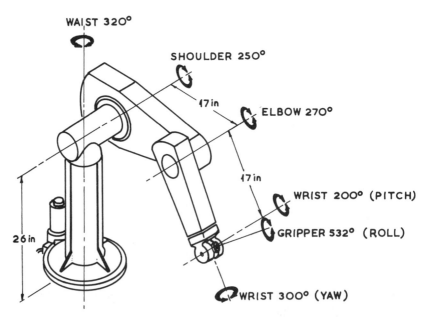

Figure 5.9   The PUMA robot manipulator.

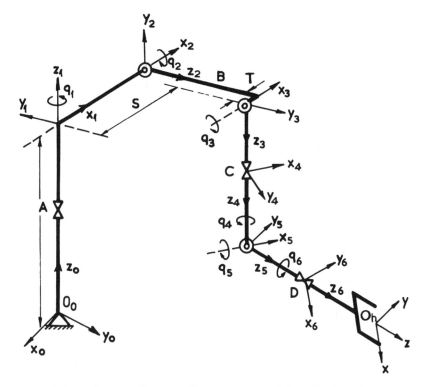

Figure 5.10 Joint coordinate systems of the PUMA robot.

$$H_{0,1} = \begin{bmatrix} \cos q_1 & -\sin q_1 & 0 & 0 \\ \sin q_1 & \cos q_1 & 0 & 0 \\ 0 & 0 & 1 & 0 \\ 0 & 0 & 0 & 1 \end{bmatrix} \tag{5.16}$$

$$H_{1,2} = \begin{bmatrix} 1 & 0 & 0 & S \\ 0 & \cos q_2 & -\sin q_2 & 0 \\ 0 & \sin q_2 & \cos q_2 & 0 \\ 0 & 0 & 0 & 1 \end{bmatrix} \tag{5.17}$$

$$H_{2,3} = \begin{bmatrix} 1 & 0 & 0 & T \\ 0 & \cos q_3 & -\sin q_3 & 0 \\ 0 & \sin q_3 & \cos q_3 & B \\ 0 & 0 & 0 & 1 \end{bmatrix} \tag{5.18}$$

$$H_{3,4} = \begin{bmatrix} \cos q_4 & -\sin q_4 & 0 & 0 \\ \sin q_4 & \cos q_4 & 0 & 0 \\ 0 & 0 & 1 & C \\ 0 & 0 & 0 & 1 \end{bmatrix} \tag{5.19}$$

$$H_{4,5} = \begin{bmatrix} 1 & 0 & 0 & 0 \\ 0 & \cos q_5 & -\sin q_5 & 0 \\ 0 & \sin q_5 & \cos q_5 & 0 \\ 0 & 0 & 0 & 1 \end{bmatrix} \tag{5.20}$$

$$H_{5,6} = \begin{bmatrix} \cos q_6 & -\sin q_6 & 0 & 0 \\ \sin q_6 & \cos q_6 & 0 & 0 \\ 0 & 0 & 1 & 0 \\ 0 & 0 & 0 & 1 \end{bmatrix} \tag{5.21}$$

We now have six linear matrices expressing the motion of each joint of the PUMA manipulator.

5.5   MANIPULATOR TRANSFORMATION GRAPH

The state of the manipulator end, i.e. the coordinate system of the last joint J6 with respect to the $(i-1)$th coordinate system, is described by $H_{i-1,6}$ and given as

$$H_{i-1,6} = H_{i-1,i} * H_{i,i+1} * H_{i+1,i+2} * \dots * H_{5,6} \tag{5.22}$$

If $i=1$ in Equation 5.22 then $H_{0,6}$ represents the state of the manipulator end with respect to the base coordinate system and the transformation matrix T_6 using the single subscript notation convention yields

$$T_6 = H_{0,6} = H_{0,1} * H_{1,2} * \dots * H_{5,6} \tag{5.23}$$

In Section 2.4 we analyzed the state of the manipulator end with respect to the base. Now we describe tool transformation and define how the manipulator fits in its environment. A pictorial representation of the coordinate transformation in Equation 5.23 is shown in Figure 5.11 (see also Figure 4.6).

The main forward path of the transformation line graph summarizes all relationships involved in the mathematical procedure in Equation 5.23 [5.3].

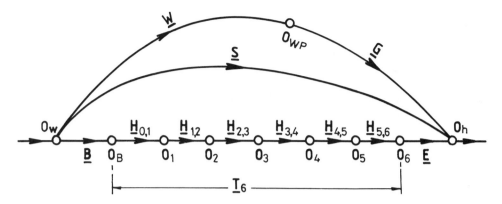

Figure 5.11 Robot manipulation transformation graph; O_0 origin of the world coordinate frame, O_B origin of the base coordinate frame, O_6 origin of the coordinate frame at the manipulator end, O_{WP} origin of actual coordinate system in workspace, O_h origin of the coordinate frame at TCP.

From the graph we directly obtain

$$B * T_6 * E = S \qquad (5.24a)$$

where $B =$ the matrix that relates the state of the manipulator base to the world coordinate system, $E =$ the matrix that represents the tool transformation and $S =$ the matrix that relates the manipulator end to the world reference system. To obtain the matrix S from the task description, see Equations 4.27 and 4.28.

Premultiplying Equation 5.24a by B^{-1} and postmultiplying by E^{-1}, then T_6 becomes

$$T_6 = B^{-1} * S * E^{-1} \qquad (5.24b)$$

5.6 CALCULATION OF T_6 FOR A SYMMETRICALLY STRUCTURED ROBOT (Hitachi assembly robot)

In this section we will calculate the manipulator transformation matrix T_6 in Equation 5.3 for the Hitachi assembly robot. We start at joint J6 and proceed backwards from the wrist flange towards the base, i.e. in the reverse direction

$$H_{5,6} = \begin{bmatrix} \cos q_6 & -\sin q_6 & 0 & 0 \\ \sin q_6 & \cos q_6 & 0 & 0 \\ 0 & 0 & 1 & B \\ 0 & 0 & 0 & 1 \end{bmatrix} \qquad (5.25)$$

The transformation within the wrist is processed as follows

$$H_{4,6} = H_{4,5} * H_{5,6}$$

$$= \begin{bmatrix} 1 & 0 & 0 & 0 \\ 0 & \cos q_5 & -\sin q_5 & 0 \\ 0 & \sin q_5 & \cos q_5 & 0 \\ 0 & 0 & 0 & 1 \end{bmatrix} * \begin{bmatrix} \cos q_6 & -\sin q_6 & 0 & 0 \\ \sin q_6 & \cos q_6 & 0 & 0 \\ 0 & 0 & 1 & D \\ 0 & 0 & 0 & 1 \end{bmatrix}$$

$$= \begin{bmatrix} \cos q_6 & -\sin q_6 & 0 & 0 \\ \cos q_5 \sin q_6 & \cos q_5 \cos q_6 & -\sin q_5 & -D \sin q_5 \\ \sin q_5 \sin q_6 & \sin q_5 \cos q_6 & \cos q_5 & D \cos q_5 \\ 0 & 0 & 0 & 1 \end{bmatrix} \tag{5.26}$$

The transformation from the elbow to the manipulator end, which involves both previous steps and a further step back to the elbow, yields

$$H_{3,6} = H_{3,4} * H_{4,6} = \begin{bmatrix} \cos q_4 & -\sin q_4 & 0 & 0 \\ \sin q_4 & \cos q_4 & 0 & 0 \\ 0 & 0 & 1 & C \\ 0 & 0 & 0 & 1 \end{bmatrix}$$

$$* \begin{bmatrix} \cos q_6 & -\sin q_6 & 0 & 0 \\ \cos q_5 \sin q_6 & \cos q_5 \cos q_6 & -\sin q_5 & -D \sin q_5 \\ \sin q_5 \sin q_6 & \sin q_5 \cos q_6 & \cos q_5 & D \cos q_5 \\ 0 & 0 & 0 & 1 \end{bmatrix} \tag{5.27a}$$

By introducing the abbreviations

$$\cos q_i = ci \text{ and } \sin q_i = si$$

then Equation 5.27a becomes

$$H_{3,6} = \begin{bmatrix} c4c6 - s4c5s6 & -c4s6 - s4c5c6 & s4s5 & Ds4s5 \\ s4c6 + c4c5s6 & -s4s6 + c4c5c6 & -c4s5 & -Dc4s5 \\ s5s6 & s5c6 & c5 & Dc5 + C \\ 0 & 0 & 0 & 1 \end{bmatrix} \tag{5.27b}$$

By appropriate grouping of the terms the computation of the matrix $H_{3,6}$ takes fifteen multiplications.

Instead of following this procedure to calculate the matrices $H_{2,6}$ and $H_{1,6}$, we proceed to derive matrices $H_{1,3}$ and $H_{0,3}$ in the forward direction. Then by multiplying $H_{0,3}$ and $H_{3,6}$ we get the overall transformation matrix T_6. In doing so we ensure some reduction in arithmetic calculations. In fact this process may be optimized by choosing a proper break in the transformation sequence and introducing a sequence of transformations partly in the forward and partly in the reverse direction to obtain more efficient computation.

The coordinate transformation from waist to elbow is performed by the following matrix manipulation:

$$H_{1,3} = H_{1,2} * H_{2,3}$$

$$= \begin{bmatrix} 1 & 0 & 0 & 0 \\ 0 & \cos q_2 & -\sin q_2 & 0 \\ 0 & \sin q_2 & \cos q_2 & 0 \\ 0 & 0 & 0 & 1 \end{bmatrix} * \begin{bmatrix} 1 & 0 & 0 & 0 \\ 0 & \cos q_3 & -\sin q_3 & 0 \\ 0 & \sin q_3 & \cos q_3 & B \\ 0 & 0 & 0 & 1 \end{bmatrix}$$

$$= \begin{bmatrix} 1 & 0 & 0 & 0 \\ 0 & \cos(q_2+q_3) & -\sin(q_2+q_3) & -B\sin q_2 \\ 0 & \sin(q_2+q_3) & \cos(q_2+q_3) & B\cos q_2 \\ 0 & 0 & 0 & 1 \end{bmatrix} \quad (5.28)$$

The resultant rotation in Equation 5.28 is the sum of the two single rotations contributed by the two adjacent bend joints.

Now the transformation matrix $H_{0,3}$ from the base to the elbow will be constituted as

$$H_{0,3} = H_{0,1} * H_{1,3} = \begin{bmatrix} \cos q_1 & -\sin q_1 & 0 & 0 \\ \sin q_1 & \cos q_1 & 0 & 0 \\ 0 & 0 & 1 & A \\ 0 & 0 & 0 & 1 \end{bmatrix}$$

$$* \begin{bmatrix} 1 & 0 & 0 & 0 \\ 0 & \cos(q_2+q_3) & -\sin(q_2+q_3) & -B\sin q_2 \\ 0 & \sin(q_2+q_3) & \cos(q_2+q_3) & B\cos q_2 \\ 0 & 0 & 0 & 1 \end{bmatrix} \quad (5.29a)$$

By introducing further trigonometrical abbreviations

$$c23 = \cos(q_2 + q_3) \text{ and } s23 = \sin(q_2 + q_3)$$

then Equation 5.29a becomes

$$H_{0,3} = \begin{bmatrix} c1 & -s1c23 & s1s23 & Bs1s2 \\ s1 & c1c23 & -c1s23 & -Bc1s2 \\ 0 & s23 & c23 & Bc2+A \\ 0 & 0 & 0 & 1 \end{bmatrix} \tag{5.29b}$$

Matrix $H_{0,3}$ can now be computed by eight multiplications.

Finally

$$H_{0,6} = T_6 = H_{0,3} * H_{3,6} \tag{5.30a}$$

and

$$T_6 = \begin{bmatrix} e_{1x} & e_{2x} & e_{3x} & k_x \\ e_{1y} & e_{2y} & e_{3y} & k_y \\ e_{1z} & e_{2z} & e_{3z} & k_z \\ 0 & 0 & 0 & 1 \end{bmatrix} \tag{5.30b}$$

Note that matrix T_6 is also obtained by a transformation given by the product of the matrices in Equations 5.27b and 5.29b.

By equating the corresponding members of the matrices $H_{0,6}$ and T_6 in Equations 5.30a and 5.30b respectively, the terms in Equation 5.30b can be readily obtained:

$$\left. \begin{aligned} e_{1x} &= c1(c4c6 - s4c5s6) - s1[c23(s4c6 + c4c5s6) - s23s5s6] \\ e_{1y} &= s1(c4c6 - s4c5s6) + c1[c23(s4c6 + c4c5s6) - s23s5s6] \\ e_{1z} &= s23(s4c6 + c4c5s6) + c23s5s6 \end{aligned} \right\} \tag{5.31a}$$

$$\left. \begin{aligned} e_{2x} &= -c1(c4s6 + s4c5c6) + s1[c23(s4s6 - c4c5c6) + s23s5c6] \\ e_{2y} &= -s1(c4s6 + s4c5c6) - c1[c23(s4s6 - c4c5c6) + s23s5c6] \\ e_{2z} &= -s23(s4s6 - c4c5c6) + c23s5c6 \end{aligned} \right\} \tag{5.31b}$$

$$
\left.
\begin{aligned}
e_{3x} &= c1s4s5 + s1(c23c4s5 + s23c5) \\
e_{3y} &= s1s4s5 - c1(c23c4s5 + s23c5) \\
e_{3z} &= -s23c4s5 + c23c5
\end{aligned}
\right\} \tag{5.31c}
$$

$$
\left.
\begin{aligned}
k_x &= D[(c1s4s5 + s1(c23c4s5 + s23c5)] + s1(Cs23 + Bs2) \\
k_y &= D[s1s4s5 - c1(c23c4s5 + s23c5)] - c1(Cs23 + Bs2) \\
k_z &= D(-s23c4s5 + c23c5) + Cc23 + Bc2 + A
\end{aligned}
\right\} \tag{5.31d}
$$

In order to evaluate the rotation submatrix $[e_1 \; e_2 \; e_3]$ in T_6 we need six sine (s1, s2, s3, s4, s5, s6) and six cosine (c1, c2, c3, c4, c5, c6) function calculations, 15 additions and 38 multiplications of appropriately grouped terms in Equation 5.31. The first column (e_1) can be obtained as the vector cross-product of the second (e_2) and third (e_3) columns. Given the joint variables $q_1, \ldots, q_6$, the state of the hand is obtained by evaluating the equation set in Equations 5.31a, b, c and d. Recognizing the relationship between Equations 5.31c and 5.31d, the last column (the k position vector) can be obtained from the third column, i.e. from Equation 5.31c, as follows

$$
k = De_3 + k' \tag{5.32}
$$

where

$$
k' = \begin{bmatrix} s1(Cs23 + Bs2) \\ -c1(Cs23 + Bs2) \\ Cc23 + Bc2 + A \end{bmatrix} \tag{5.33}
$$

This contributes another six multiplications in addition to the number of arithmetic computations summarized above.

5.7 CALCULATION OF T_6 FOR AN ASYMMETRICALLY STRUCTURED ROBOT (Unimation PUMA 600)

We will now deal with the kinematic equations for the PUMA robot. The joint axes of the wrist intersect at one point.

Orientation of the three wrist axes is instituted by matrices performed in reverse sequence: $H_{5,6} \ldots H_{4,6}$ etc.

$$
H_{4,6} = H_{4,5} * H_{5,6} = \begin{bmatrix} c6 & -s6 & 0 & 0 \\ c5s6 & c5c6 & -s5 & 0 \\ s5s6 & s5c6 & c5 & 0 \\ 0 & 0 & 0 & 1 \end{bmatrix} \tag{5.34}
$$

$$H_{3,6} = H_{3,4} * H_{4,6}$$

$$
= \begin{bmatrix}
c4c6 - s4c5s6 & -c4s6 - s4c5c6 & s4s5 & 0 \\
s4c6 + c4c5s6 & -s4s6 + c4c5c6 & -c4s5 & 0 \\
s5s6 & s5c6 & c5 & C \\
0 & 0 & 0 & 1
\end{bmatrix}
\tag{5.35}
$$

The coordinate transformation in the reverse direction is implemented through to the base:

$$H_{2,6} = H_{2,3} * H_{3,6} =$$

$$
\begin{bmatrix}
c4c6 - s4c5s6 & -c4s6 - s4c5c6 & s4s5 & T \\
c3(s4c6 + c4c5s6) - s3s5s6 & c3(-s4s6 + c4c5c6) - s3s5c6 & -c3c4s5 - s3s5 & -Cs3 \\
s3(s4c6 + c4c5s6) + c3s5s6 & s3(-s4s6 + c4c5c6) + c3s5c6 & -s23c4s5 + c3c5 & Cs3 + B \\
0 & 0 & 0 & 1
\end{bmatrix}
$$
$$\tag{5.36}$$

The joint variables q_2 and q_3 operate in parallel because they are adjacent transverse axes. Therefore we can simply add them up in the rotation submatrix:

$$
H_{1,3} = H_{1,2} * H_{2,3} = \begin{bmatrix}
1 & 0 & 0 & T+S \\
0 & c23 & -s23 & -Bs2 \\
0 & s23 & c23 & Bc2 \\
0 & 0 & 0 & 1
\end{bmatrix}
\tag{5.37}
$$

$$H_{1,6} = H_{1,3} * H_{3,6} \cdots =$$

$$
\begin{bmatrix}
c4c6 - s4c5s6 & -c4s6 - s4c5c6 & s4s5 & T+S \\
c23(s4c6 + c4c5s6) - s23s5s6 & c23(-s4s6 + c4c5c6) - s23s5c6 & -c23c4s5 - s23s5 & -Cs23 - Bs2 \\
s23(s4c6 + c4c5s6) + c23s5s6 & s23(-s4s6 + c4c5c6) + c23s5c6 & -s23c4s5 + c23c5 & Cs23 + Bc2 \\
0 & 0 & 0 & 1
\end{bmatrix}
$$

$$\tag{5.38}$$

Finally

$$T_6 = H_{0,1} * H_{1,6} = \begin{bmatrix} e_{1x} & e_{2x} & e_{3x} & k_x \\ e_{1y} & e_{2y} & e_{3y} & k_y \\ e_{1z} & e_{2z} & e_{3z} & k_z \\ 0 & 0 & 0 & 1 \end{bmatrix} \qquad (5.39)$$

where

$$\left. \begin{aligned} e_{1x} &= c1(c4c6 - s4c5s6) - s1[c23(s4c6 + c4c5s6) - s23s5s6] \\ e_{1y} &= s1(c4c6 - s4c5s6) + c1[c23(s4c6 + c4c5s6) - s23s5s6] \\ e_{1z} &= s23(s4c6 + c4c5s6) + c23s5s6 \end{aligned} \right\} \qquad (5.40a)$$

$$\left. \begin{aligned} e_{2x} &= -c1(c4s6 + s4c5c6) + s1[c23(s4s6 - c4c5c6) + s23s5c6] \\ e_{2y} &= -s1(c4s6 + s4c5c6) - c1[c23(s4s6 - c4c5c6) + s23s5c6] \\ e_{2z} &= -s23(s4s6 - c4c5c6) + c23s5c6 \end{aligned} \right\} \qquad (5.40b)$$

$$\left. \begin{aligned} e_{3x} &- c1s4s5 + s1(c23c4s5 + s23c5) \\ e_{3y} &= s1s4s5 - c1(c23c4s5 + s23c5) \\ e_{3z} &= -s23c4s5 + c23c5 \end{aligned} \right\} \qquad (5.40c)$$

$$\left. \begin{aligned} k_x &= c1(T + S) + s1(Cs23 + Bs2) \\ k_y &= s1(T + S) - c1(Cc23 + Bs2) \\ k_z &= Cc23 + Bc2 \end{aligned} \right\} \qquad (5.40d)$$

5.8 *ILLUSTRATIVE EXAMPLE 12*
DERIVATION OF THE KINEMATIC MODEL OF A SCARA ROBOT (Dainichi–Sykes Daros PT300H)

Specification

Daros PT300H is a Dainichi–Sykes robot horizontally operating with a vertical stroke joint at the end. Recently this type of robot structure has been classified as a SCARA type robot (see Section 1.5).

Daros PT300H is a four degrees of freedom robot system. Joints J1, J2 and J4 are of the swivel type and J3 is a prismatic joint. So the robot is of RRPR type. There are two horizontal offsets shown as *B* and *D*.

(a)

Model		PT300H	
Type		Jointed Arm	
Load carrying capability		5 kg	
Number of axes		4	
	Axis	Momentary max speed	Operating range
Main axes	Shoulder arm (X)	90°/s	±90°
	Elbow arm (Z)	90°/s	+116° −144°
Wrist axis	Horizontal rotation (β)	90°/s	±105°
	Vertical stroke (Pneumatic)	100 mm/s	100 mm
Drive system		DC Servo Motor (Optional DC Servo vertical stroke)	
Axis positioning repeatability		±0.1 mm	
Balancing mechanism		—	
Power requirements	Electric	3 phase AC380/415 V, 50/60 Hz, 1.0 kVA	
	Pneumatic	6.5 kg/cm²G, 40 l/min nominal	
Weight		125 kg	

(b)

(c)

Figure 5.12 Dainichi–Sykes Daros PT300H SCARA robot. (a) Photograph of the robot. (b) Performance specification. (c) A working envelope and dimensions.

Objectives

The aim is to describe the end effector's position and orientation by using direct transformation. Start from the base and then proceed towards the end effector. Provide the robot parameters and the derivation of the transformation matrices to the joints of the robot arm.

Solutions

To start with we will set up an appropriately oriented coordinate frame at each joint (see Figure 5.13).

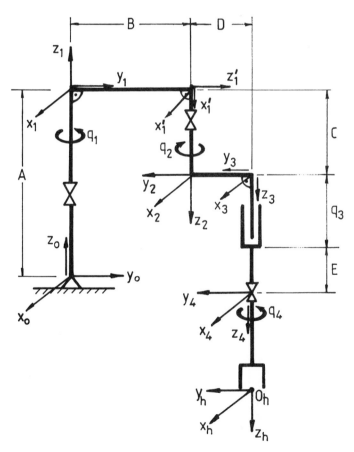

Figure 5.13 Coordinate frames of the joints of the Dainichi–Sykes Daros PT300H SCARA robot.

To start with, there is a rotation q_1 about the z_0 axis and a translation A along the same z_0:

$$H_{0,1} = \begin{bmatrix} \cos q_1 & -\sin q_1 & 0 & 0 \\ \sin q_1 & \cos q_1 & 0 & 0 \\ 0 & 0 & 1 & A \\ 0 & 0 & 0 & 1 \end{bmatrix} \qquad (5.41)$$

The $H_{1,2}$ matrix implements the transformation as follows:

(1) A translation B along y_1.
(2) A rotation $+90°$ about x_1.
(3) A translation C along y'_1.
(4) A rotation $+90°$ about x_2.
(5) A rotation q_2 about z_2.

thus

$$H_{1,2} = \begin{bmatrix} 1 & 0 & 0 & 0 \\ 0 & 0 & 1 & B \\ 0 & -1 & 0 & 0 \\ 0 & 0 & 0 & 1 \end{bmatrix} * \begin{bmatrix} 1 & 0 & 0 & 0 \\ 0 & 0 & 1 & C \\ 0 & -1 & 0 & 0 \\ 0 & 0 & 0 & 1 \end{bmatrix}$$

$$* \begin{bmatrix} \cos q_2 & -\sin q_2 & 0 & 0 \\ \sin q_2 & \cos q_2 & 0 & 0 \\ 0 & 0 & 1 & C \\ 0 & 0 & 0 & 1 \end{bmatrix}$$

$$= \begin{bmatrix} \cos q_2 & -\sin q_2 & 0 & B \\ -\sin q_2 & -\cos q_2 & 0 & 0 \\ 0 & 0 & -1 & -C \\ 0 & 0 & 0 & 1 \end{bmatrix} \qquad (5.42)$$

The $H_{2,3}$ matrix represents a translation D along the $-y_2$ axis and a varying translation q_3 along the z_3 axis

$$H_{2,3} = \begin{bmatrix} 1 & 0 & 0 & 0 \\ 0 & 1 & 0 & -D \\ 0 & 0 & 1 & q_3 \\ 0 & 0 & 0 & 1 \end{bmatrix} \tag{5.43}$$

The $H_{3,4}$ matrix is composed of a rotation q_4 about the z_3 and a translation E along the same axis.

$$H_{3,4} = \begin{bmatrix} \cos q_4 & -\sin q_4 & 0 & 0 \\ \sin q_4 & \cos q_4 & 0 & 0 \\ 0 & 0 & 1 & E \\ 0 & 0 & 0 & 1 \end{bmatrix} \tag{5.44}$$

The total transformation results in matrix

$$H_{0,4} = H_{0,1} * H_{1,2} * H_{2,3} * H_{3,4} \tag{5.45}$$

By introducing the usual abbreviations $\cos q_1 = c1$ and $\sin q_1 = s1$, and then expanding Equation 5.45, we obtain

$$H_{0,4} = \begin{bmatrix} c1 & -s1 & 0 & 0 \\ s1 & c1 & 0 & 0 \\ 0 & 0 & 1 & A \\ 0 & 0 & 0 & 1 \end{bmatrix} * \begin{bmatrix} c2 & -s2 & 0 & B \\ -s2 & -c2 & 0 & B \\ 0 & 0 & -1 & -C \\ 0 & 0 & 0 & 1 \end{bmatrix}$$

$$* \begin{bmatrix} 1 & 0 & 0 & 0 \\ 0 & 1 & 0 & -D \\ 0 & 0 & 1 & q_3 \\ 0 & 0 & 0 & 1 \end{bmatrix} * \begin{bmatrix} c4 & -s4 & 0 & 0 \\ s4 & c4 & 0 & 0 \\ 0 & 0 & 1 & E \\ 0 & 0 & 0 & 1 \end{bmatrix}$$

$$= \begin{bmatrix} c12c4 + s12s4 & -c12s4 + s12c4 & 0 & -Ds12 - Bs1 \\ s12c4 - c12s4 & -s12s4 - c12c4 & 0 & Dc12 + Bc1 \\ 0 & 0 & -1 & -E - q_3 + A - C \\ 0 & 0 & 0 & 1 \end{bmatrix}$$

where

$$c12 = \cos(q_1 - q_2) \quad \text{and} \quad s12 = \sin(q_1 - q_2)$$

$$H_{0,4} = \begin{bmatrix} c124 & -s124 & 0 & -Ds12 - Bs1 \\ s124 & c124 & 0 & Dc12 + Bc1 \\ 0 & 0 & -1 & (A - C - E) - q_3 \\ 0 & 0 & 0 & 1 \end{bmatrix} \tag{5.46}$$

where

$$c124 = \cos(q_1 - q_2 - q_4) \quad \text{and} \quad s124 = \sin(q_1 - q_2 - q_4)$$

To check the result, put all rotations and q_3 equal to zero, then the resultant matrix is readily obtainable from Figure 5.13.

$$\begin{bmatrix} 1 & 0 & 0 & 0 \\ 0 & -1 & 0 & B + D \\ 0 & 0 & -1 & A - C - E \\ 0 & 0 & 0 & 1 \end{bmatrix} \tag{5.47}$$

In Figure 5.13 in accordance with Equation 5.47 $X_0 \parallel X_h$, $T_0 \parallel T_h$, $Z_0 \parallel Z_h$, the offset in the X_0 direction is zero, in the T_0 direction $B + D$, and in the Z_0 direction $A - C - E$.

5.9   COMPUTATIONAL AND PROGRAMMING CONSIDERATIONS

Advanced robot control systems are able to calculate the base-to-hand transformation in real time, i.e. while the robot has already started to execute a task. The involvement of online calculation naturally requires that the number of arithmetic operations should be reduced to a minimum, and the arithmetic complexity be optimized.

As we have already seen, the elements of T_6 consist of polynomials of sines and cosines of the six joint angles. Each polynomial consists of sine and cosine functions with at least one joint variable. In the position vector, the fourth column in the transformation matrices, the link and offset constants in Sections 5.6, 5.7 and 5.8 (A, B, C, D, S and T) are multiplied by the sines and cosines of the joint variables. In order to establish an optimally operating computing method, various options for performing the developed manipulator transformation matrices [5.4] must be considered which can then be programmed into the computer assisted controller.

The primary concern is the number of multiplications, since additions and subtractions require an order of magnitude less of computation time. The time requirement for transcendental function calls is often very low, since sine and cosine functions can be obtained from preprogrammed look-up tables set in computer memories.

In order to appreciate the computational involvement, for instance with the Hitachi assembly robot, a comprehensive summary follows:

(a) We could simply construct the six 4×4 joint transformation matrices related to the six degrees of freedom and multiply them together. This would need

$$(4 \times 4 \times 4)(6-1) = 320 \text{ multiplications}$$

(b) Since the fourth row in each component matrix is $(0\ 0\ 0\ 1)$ we could simply carry it in the resultant matrix. Thus the arithmetic is reduced to

$$(4 \times 3 \times 3)(6-1) = 180 \text{ multiplications}$$

(c) We know that each component matrix specifies only a single rotation about a coordinate axis which changes only two coordinates at the same time (either x and y or y and z) and also may consist of a shift towards x or z. Thus the general matrix multiplication may be replaced by mere rotation and shift operators.

This requires

$$(4 \times 2 \times 2)(6-1) = 80 \text{ multiplications}$$

(d) We note that in Equation 5.27b the result of the last three rotations which correspond to the Euler angles can be calculated by 15 multiplications. Taking into account the considerations in (c) we have to carry out

$$15 + (4 \times 2 \times 2)(4-1) = 63 \text{ multiplications}$$

(e) If we consider that q_2 and q_3 are simply added, i.e. $H_{1,3}$ is produced with only two multiplications in the offset vector, we get

$$15 + (4 \times 2 \times 2)(3-1) + 2 = 49 \text{ multiplications}$$

(f) If we calculate $H_{0,3}$ first by using eight multiplications as indicated in Equation 5.29 and then we go on multiplying, one by one, we get

$$8 + [(2 \times 2 + 1) \times 3] (4 - 1) = 53 \text{ multiplications}$$

(g) And finally, if we calculate $H_{0,3}$ and $H_{3,6}$ separately as given in Equations 5.27 and 5.29 and then multiply them together, we end up with

$$8 + 15 + (3 \times 4 \times 3) = 59 \text{ multiplications}$$

(h) By symbolic calculation of T_6 and appropriate grouping of its elements we could solve the task with 38 multiplications as seen in the set of Equations 5.31a–5.31d.

The method described under (h) requires the least number of multiplications. The algorithm described under (e) has, however, the advantage that it calculates $H_{3,6}$ separately and the rotation part, a 3×3 upper left submatrix of $H_{3,6}$, has the same form as the matrix describing the hand orientation. To see this point compare Equation 5.27b with Equation 3.5 in Sections 5.6 and 3.1 respectively. Thus we can use the same subroutine to compute both transformations.

5.10   REFERENCES

[5.1]   Hartenberg, R. S. and Denavit, J. *Kinematic Synthesis of Linkages*, Chapter 12, McGraw-Hill, 1964.

[5.2]   Konstantinov, M. S. Inertia forces of robots and manipulators, *Conf. on Remotely Manned Syst.(RMS): Technology and Application*, 2nd Proc., *Los Angeles, Calif.*, June 9–11, 1975, pp. 31–2. Publ. by NASA, Washington, DC, 1975.

[5.3]   Paul, R. P. *Robot Manipulators: Mathematics, Programming and Control*, The MIT Press, 1981.

[5.4]   Horn, B. K. P. and Inove, H. Kinematics of the MIT-Al-VICARM manipulator, *Working Paper 69*, Massachusetts Inst. of Technology, Artificial Intelligence Lab., May 1974.

6

INVERSE MODELING OF ROBOT MANIPULATORS

6.1   INVERSE MODELING OF MANIPULATOR GEOMETRY

The calculation of joint variables for a given state of the end effector anywhere in the workspace is the key problem in computer controlled robot manipulation. The direct modeling method involves calculating the end effector's states from joint angles. But for many robot applications it is necessary to calculate the joint angles from the states of the end effector. For distinction from the direct procedure, the scheme is called inverse task modeling. In this chapter we shall outline some methods leading to these inverse task solutions.

Having developed the model for characterizing the state of the end effector in Section 5.1 and having described the robot geometry in mathematical forms in Section 5.2, we will now deal with the formulation of mathematical models of the industrial robot's inverse kinematic structure. The most common task in manipulator control is where to drive the manipulator end effector in the cartesian workspace. We know this from the task description but we do not know yet how to do it. In other words we have to know what are the manipulator's joint variables (angles and/or offsets) related to a given state (position and orientation) of the end effector.

While there is only one end effector position and orientation corresponding to a given set of joint variables, there may be a large number of possible manipulator joint-link configurations represented by different sets of realizable joint variables, all of which could place the end effector in the same desired position and orientation. We usually evaluate all sets of solutions but finally use only one set of them. As we shall see, there is no problem obtaining the solutions, the problem is how to select the optimal set from the available joint-link configurations.

If the manipulator geometry is very simple, that is the robot has only a few controllable joints, or some of the links are of zero length, a direct trigonometric solution is readily obtainable [6.1]. However, manipulators with six degrees of freedom are often much more complex and therefore the direct trigonometric method would be unable to provide a solution.

In Section 6.2 we present a method for solving the inverse equations, based on homogeneous transformation matrices, from which the solution can explicitly be deduced in the case of a wide range of manipulators. Practically all commercially available manipulators can be considered as kinematically simple and the methods presented in the following are directly applicable to them. The existence of an explicit solution to the kinematic equations is very important from the point of view of deciding which type of computer control is most suitable for a particular manipulator (see Chapter 7).

Let us consider first what happens if we tackle the problem in a straightforward manner. Since each element in the manipulator transformation matrix T_6 (see Equation 5.27) is a polynomial in sine and cosine functions of the joint variables, and in constant offsets, in the case of asymmetrical robots, one would naturally decide to solve the set of twelve polynomial equations, six for sines and six for cosines. This can be done by the successive elimination of variables, analogous to the method employed for the solution of linear equations. The successive elimination of unknowns would finally lead to a single high order equation having only one unknown variable.

We could still try to solve the problem without any geometrical intuition. There is an alternative scheme which interpolates from a stored set of configurations (joint variable sets) combined with iteration and gradient methods (optimization or searching routines). We could obtain the solution by repeatedly evaluating the manipulator transformation matrices so providing the values of joint variables. But this is quite a slow process for real-time computation and would need an extremely large amount of memory to map the complex joint-link configurations to yield sufficient resolution. An example of this approach is presented by Albus [6.2].

6.2   THE METHOD OF INVERSE TRANSFORMATION MATRICES

The method for inverse transformation applicable to most manipulators is based on trigonometry and uses homogeneous transformation which yields equations in terms of sine and cosine of the joint angles [6.3]. These equations are then combined using the arctan functions of two arguments to avoid the problem of quadrant ambiguities.

To start with we note that the manipulator transformation matrix T_6 is already known to us from the task description as it was constituted in Chapter 4, and it is equal to the product of six elementary transformation matrices given by

$$T_6 = H_{0,1} * H_{1,2} * \ldots * H_{5,6} \tag{6.1}$$

Premultiplying Equation 6.1 successively by the inverse of the matrix $H_{i-1,i}$, we obtain five discrete matrix equations, namely

$$H_{0,1}^{-1} * T_6 = H_{1,6} \tag{6.2}$$

$$H_{1,2}^{-1} * H_{0,1}^{-1} * T_6 = H_{2,6} \tag{6.3}$$

$$H_{2,3}^{-1} * H_{1,2}^{-1} * H_{0,1}^{-1} * T_6 = H_{3,6} \tag{6.4}$$

$$H_{3,4}^{-1} * H_{2,3}^{-1} * H_{1,2}^{-1} * H_{0,1}^{-1} * T_6 = H_{4,6} \tag{6.5}$$

$$H_{4,5}^{-1} * H_{3,4}^{-1} * H_{2,3}^{-1} * H_{1,2}^{-1} * H_{0,1}^{-1} * T_6 = H_{5,6} \tag{6.6}$$

We find matrices on the left-hand sides of these equations which are the function of the elements of T_6 and the 1st to $(i-1)$th joint variables. The matrix elements on the right-hand sides are either zeros, constants or functions of the ith to 6th joint variables. Thus, by equating matrix elements one by one, we obtain twelve equations from each matrix equation, i.e. one for each component of the 3×1 column vectors e_1, e_2, e_3 and k.

The solution starts by developing both sides of Equation 6.2. Provided that q_1 represents a revolute joint variable, the left-hand side of Equation 6.2 will correspond to one of the following two types of relationship.

(1) If the 1st joint of a robot is of the swivel type turning about the z axis like the waist of the PUMA robot then Equation 6.2 results in the following expanded form

$$H_{0,1}^{-1} * T_6 = \begin{bmatrix} c1 & s1 & 0 & 0 \\ -s1 & c1 & 0 & 0 \\ 0 & 0 & 1 & 0 \\ 0 & 0 & 0 & 1 \end{bmatrix} * \begin{bmatrix} e_{1x} & e_{2x} & e_{3x} & k_x \\ e_{1y} & e_{2y} & e_{3y} & k_y \\ e_{1z} & e_{2z} & e_{3z} & k_z \\ 0 & 0 & 0 & 1 \end{bmatrix}$$

$$= \begin{bmatrix} c1e_{1x}+s1e_{1y} & c1e_{2x}+s1e_{2y} & c1e_{3x}+s1e_{3y} & c1k_x+s1k_y \\ -s1e_{1x}+c1e_{1y} & -s1e_{2x}+c1e_{2y} & -s1e_{3x}+c1e_{3y} & -s1k_x+c1k_y \\ e_{1z} & e_{2z} & e_{3z} & k_z \\ 0 & 0 & 0 & 1 \end{bmatrix} \quad (6.7)$$

(2) If the 1st joint of a robot is of the bend type, turning about the x axis, then Equation 6.2 results in the following expanded form

$$H_{0,1}^{-1} * T_6 = \begin{bmatrix} 1 & 0 & 0 & 0 \\ 0 & c1 & s1 & 0 \\ 0 & -s1 & c1 & 0 \\ 0 & 0 & 0 & 1 \end{bmatrix} * \begin{bmatrix} e_{1x} & e_{2x} & e_{3x} & k_x \\ e_{1y} & e_{2y} & e_{3y} & k_y \\ e_{1z} & e_{2z} & e_{3z} & k_z \\ 0 & 0 & 0 & 1 \end{bmatrix}$$

$$= \begin{bmatrix} e_{1x} & e_{2x} & e_{3x} & k_x \\ c1e_{1y}+s1e_{1z} & c1e_{2y}+s1e_{2z} & c1e_{3y}+s1e_{3z} & c1k_y+s1k_z \\ -s1e_{1y}+c1e_{1z} & -s1e_{2y}+c1e_{2z} & -s1e_{3y}+c1e_{3z} & -s1k_y+c1k_z \\ 0 & 0 & 0 & 1 \end{bmatrix} \quad (6.8)$$

We first calculate the angle q_1 by equating and combining the corresponding elements of either Equation 6.7 or 6.8 with $H_{1,6}$, depending upon the joint type we are dealing with. In so doing, we have to select those elements of matrix $H_{1,6}$ which contain the least number of unknowns. This method has first been developed for the Stanford manipulator having one prismatic and five revolute joints discussed in detail by Paul [6.3].

6.3   SOLUTION OF THE INVERSE TASK OF A REVOLUTE ROBOT (Unimation PUMA 600)

In this section we will derive the solution of the inverse task, step by step, for the PUMA 600 robot, which has six joints and two offsets.

(1) Determination of q_1

We first consider a swivel joint configuration. We premultiply T_6 by the inverse of the matrix $H_{0,1}$ to get Equation 6.2, namely

$$H_{0,1}^{-1} * T_6 = H_{1,6}$$

where matrix $H_{1,6}$ is already defined by Equation 5.38 and all its elements are functions of q_2, $q_{2,3}$, q_4, q_5 and q_6 except the matrix element 1,4, which represents both offsets T and S in the robot structure. Thus equating the 1, 4 element of Equation 6.7 to the corresponding element in Equation 5.38, we obtain

$$c1k_x + s1k_y = T + S \qquad (6.9)$$

Let us now introduce the following substitutions, bearing in mind that vector k (k_x, k_y, k_z) yields the position of the end effector in a cartesian coordinate system and r is the projection of k in the (x, y) plane

$$k_x = r \cos \varphi \qquad (6.10)$$

$$k_y = r \sin \varphi \qquad (6.11)$$

where

$$r = (k_x^2 + k_y^2)^{1/2} \qquad (6.12)$$

and

$$\varphi = \arctan \frac{k_y}{k_x} \qquad (6.13)$$

The arctan function uses the sign of the numerator and of the denominator to determine the correct quadrant for the resulting angle and is defined over the range

$$-180° \leq \varphi < 180°$$

By substituting Equations 6.10 and 6.11 into Equation 6.9 we obtain

$$c1 \cos \varphi + s1 \sin \varphi = \frac{T+S}{r} \qquad (6.14)$$

where

$$0 \leq \frac{T+S}{r} < 1 \qquad (6.15)$$

Then by applying the trigonometrical relation in Equation 2.A.4, Equation (6.14) is reduced to

$$\cos(\varphi - q_1) = \frac{T+S}{r} \qquad (6.16)$$

where

$$0° \leq (\varphi - q_1) < 180°$$

We may obtain the sine term as

$$\sin(\varphi - q_1) = \pm \sqrt{1 - \left(\frac{T+S}{r}\right)^2} \qquad (6.17)$$

where the plus sign corresponds to a right-hand and the minus sign to the left-hand configuration of the manipulator.

From Equations 6.16 and 6.17 we get

$$\tan(\varphi - q_1) = \frac{\pm\sqrt{r^2 - (T+S)^2}}{T+S} \qquad (6.18)$$

Thus by substituting for φ from Equation 6.13, we obtain

$$q_1 = \arctan \frac{k_y}{k_x} - \arctan\left\{\frac{\pm\sqrt{k_x^2 + k_y^2 - (T+S)^2}}{T+S}\right\} \qquad (6.19)$$

Note that q_1 represents the swivel rotation of the waist of the robot.

(2) Determination of q_2 and q_3

From a knowledge of q_1, the left-hand side of Equation 6.2 can be determined. Whenever we have the left-hand side of Equations 6.2–6.6 we can look for the corresponding matrix elements on the right-hand side which are a function of the coordinates of the individual joint.

In the case of any manipulator having adjacent parallel transverse axes, as the PUMA 600 has, the T_6 matrix is expressed in terms of sums and/or differences of the angles relative to the parallel axes. While solving these equations, the sums and/or differences must be found before the angles themselves can be determined. The sums and/or differences of angles can be found by taking the sum of two equations. This is the way to solve for the values of q_2 and q_3. Let the first equation solved be Equation 6.2. The 2, 4 and 3, 4 elements of the matrix $H_{1,6}$ in Equation 6.2 are retrieved from Equations 5.38 and 6.7 as

$$-Cs23 - Bs2 = -s1k_x + c1k_y \tag{6.20}$$

$$Cc23 + Bc2 = k_z \tag{6.21}$$

after squaring, adding and simplifying we obtain

$$C^2 + B^2 + 2BCc3 = s1^2k_x^2 + c1^2k_y^2 + c1^2k_y^2 + k_z^2 - 2k_xk_yc1s1 \tag{6.22}$$

and

$$c3 = \frac{s1^2k_x^2 + c1^2k_y^2 + k_z^2 - 2k_xk_yc1s1 - B^2 - C^2}{2BC} \tag{6.23}$$

$$s3 = \pm(1 - c3^2)^{1/2} \tag{6.24}$$

and finally

$$q_3 = \arctan\left[\frac{\pm(1 - c3^2)^{1/2}}{c3}\right] \tag{6.25}$$

As the calculation of q_2 is not available from Equation 6.2, the second equation is required to continue the calculation procedure. Therefore we go on one step further and evaluate Equation 6.3, namely

$$H_{1,2}^{-1} * H_{0,1}^{-1} * T_6 = H_{2,6} \tag{6.3}$$

The left-hand side will be derived in terms of the column vectors a_i, $i = 1, 2, 3$

and k, as

$$H_{1,2}^{-1} * H_{0,1}^{-1} * T_6 = [a_1 \; a_2 \; a_3 \; a_k] \tag{6.26}$$

where each of the column vectors a_i is represented by

$$
\begin{aligned}
a_1 &= [a_{1X} \quad a_{1Y} \quad a_{1Z} \quad 0]^T \\
a_2 &= [a_{2X} \quad a_{2Y} \quad a_{2Z} \quad 0]^T \\
a_3 &= [a_{3X} \quad a_{3Y} \quad a_{3Z} \quad 0]^T \\
a_k &= [a_{kX} \quad a_{kY} \quad a_{kZ} \quad 1]^T
\end{aligned}
\tag{6.27}
$$

This means that the matrix has three types of element in each column. We premultiply Equation 6.2 by the inverse of $H_{1,2}$ and from that product we deduce at first only the components of the a_1 vector in Equation 6.26, namely

$$
a_1 =
\begin{bmatrix}
a_{1X} \\
a_{1Y} \\
a_{1Z} \\
0
\end{bmatrix}
=
\begin{bmatrix}
c1e_{1x} + s1e_{1y} \\
c2(-s1e_{1x} + c1e_{1y}) + s2e_{1z} \\
-s2(-s1e_{1x} + c1e_{1y}) + c2e_{1z} \\
0
\end{bmatrix}
\tag{6.28}
$$

For the sake of completeness the other vectors a_2, a_3 and a_k in Equation 6.27 can be derived by a similar procedure to that outlined above. (For further details see Appendix 6.A.)

The terms of a_1 in Equation 6.28 and those of a_2, a_3 and a_k in Appendix 6.A are not applicable for deriving q_2, because it is not available on its own when equating the corresponding terms of matrix $H_{2,6}$ in Equation 5.36. There, only q_3 stands above in any of its matrix elements. From the value of q_3 which we have already found in Equation 6.25, we may use the corresponding terms for keeping check on the value of q_2. So for the derivation of q_2 we continue the process and proceed to the next equation, i.e. to Equation 6.4.

$$H_{2,3}^{-1} * H_{1,2}^{-1} * H_{0,1}^{-1} * T_6 = H_{3,6}$$

The left-hand side, similarly to Equation 6.26, will be derived in terms of column vectors as

$$H_{2,3}^{-1} * H_{1,2}^{-1} * H_{0,1}^{-1} * T_6 = [b_1 \; b_2 \; b_3 \; b_k] \tag{6.29}$$

where the column vectors on the right-hand side are expressed in expanded form by

$$\begin{aligned}
\boldsymbol{b}_1 &= [b_{1X} \quad b_{1Y} \quad b_{1Z} \quad 0]^T \\
\boldsymbol{b}_2 &= [b_{2X} \quad b_{2Y} \quad b_{2Z} \quad 0]^T \\
\boldsymbol{b}_3 &= [b_{3X} \quad b_{3Y} \quad b_{3Z} \quad 0]^T \\
\boldsymbol{b}_k &= [b_{kX} \quad b_{kY} \quad b_{kZ} \quad 1]^T
\end{aligned} \tag{6.30}$$

The aim now is to derive the resultant matrix of the left-hand side of Equation 6.29 in order to find a single term in q_2 among its elements. We can see that there is no term where q_2 would stand on its own; we have the chance, however, to find $(q_2 + q_3)$ denoted by q_{23}. Since q_{23} is obtainable, q_{23} and q_3 uniquely determine q_2. Similarly to Equation 6.28 we obtain the following types of matrix elements:

$$\boldsymbol{b}_1 = \begin{bmatrix} b_{1X} \\ b_{1Y} \\ b_{1Z} \\ 0 \end{bmatrix} = \begin{bmatrix} c1e_{1x} + s1e_{1y} \\ c23(-s1e_{1x} + c1e_{1y}) + s23e_{1z} \\ -s23(-s1e_{1x} + c1e_{1y}) + c23e_{1z} \\ 0 \end{bmatrix} \tag{6.31}$$

The other column vectors $\boldsymbol{b}_2$, $\boldsymbol{b}_3$ and $\boldsymbol{b}_k$ in Equation 6.29 can be derived by the same procedure as above (see Appendix 6.B).

We now can equate the 2,4 and 3,4 elements from the resultant matrix of Equation 6.4 with the elements in the same position as in Equation 5.35 and then we obtain the following relationships:

$$c23w + s23e_{1z} = 0 \tag{6.32}$$

$$-s23w + c23e_{1z} = C \tag{6.33}$$

where

$$w = -s1k_x + c1k_y \tag{6.34}$$

w is readily obtainable as far as q_1 and both k_x and k_y have already been derived in Equations 6.19, 6.10 and 6.11 respectively. By solving Equations 6.32 and 6.33 simultaneously, we get

$$c23 = C \frac{k_z}{k_z^2 - w^2} \tag{6.35}$$

$$s23 = -C \frac{w}{k_z^2 - w^2} \tag{6.36}$$

$$q_{23} = q_2 + q_3 = \arctan\left(\frac{s1k_x - c1k_y}{k_z}\right) \tag{6.37}$$

and we already have q_3 from Equation 6.25, so

$$q_2 = q_{23} - q_3 \tag{6.38}$$

(3) Derivation of q_4

To derive q_4 we go on to use Equation 6.4. We have expanded its left-hand side in Equation 6.29 and we can check the right-hand side for functions with a single variable in q_4. The 1,3 and 2,3 elements in Equation 5.35 provide us with the equations for sine and cosine in q_4 if s5 is not zero (see vector b_3 in Equation 6.B.3, Appendix 6.B).

$$s4s5 = c1e_{3x} + s1e_{3y} \tag{6.39}$$

$$-c4s5 = c23(-s1e_{3x} + c1e_{3y}) + s23e_{3z} \tag{6.40}$$

and

$$q_4' = \arctan \frac{c1e_{3x} + s1e_{3y}}{c23(s1e_{3x} - c1e_{3y}) - s23e_{3z}} \tag{6.41}$$

For $q_5 > 0$, $q_4 = q_4'$ and for $q_5 < 0$, $q_4 = q_4' + 180°$ \hfill (6.42a)

If s5=0 then $q_5 = 0$ and joints J4 and J6 are aligned (degenerate configuration). In this case only the sum of q_4 and q_6 is significant and therefore we may choose one of them arbitrarily.

For $q_5 = 0$, then we can choose $q_4 = q_4'$ and $q_6 = 0$ \hfill (6.42b)

(4) Determination of q_5 and q_6

We will now evaluate matrix $H_{4,6}$ in Equation 6.5, namely

$$H_{3,4}^{-1} * H_{2,3}^{-1} * H_{1,2}^{-1} * H_{0,1}^{-1} * T_6 = H_{4,6} \tag{6.5}$$

The left-hand side, similarly to Equation 6.29, will be derived in terms of column vectors as

$$H_{3,4}^{-1} * H_{2,3}^{-1} * H_{1,2}^{-1} * H_{0,1}^{-1} * T_6 = [c_1 \; c_2 \; c_3 \; c_k] \tag{6.43}$$

where the right-hand side is expressed in expanded form as

$$c_1 = [c_{1X} \quad c_{1Y} \quad c_{1Z} \quad 0]^T$$
$$c_2 = [c_{2X} \quad c_{2Y} \quad c_{2Z} \quad 0]^T$$
$$c_3 = [c_{3X} \quad c_{3Y} \quad c_{3Z} \quad 0]^T \tag{6.44}$$
$$c_k = [c_{kX} \quad c_{kY} \quad c_{kZ} \quad 1]^T$$

After we have expanded the left-hand side of Equation 6.5 (see Appendix 6.C), we then check the right-hand side of Equation 6.5 for functions with a single variable in q_5. Since both sine and cosine functions of q_5 are available in a single term in Equation 5.34, we get a unique value for this angle. By equating the 2,3 and 3,3 elements on both sides we can derive q_5 (see Equation 6.C.3, Appendix 6.C). The required elements are

$$s5 = -s4(c1e_{3x} + s1e_{3y}) - c4[c23(-s1e_{3x} + c1e_{3y}) + s23e_{3z}] \tag{6.45}$$

$$c5 = -s23(-s1e_{3x} + c1e_{3y}) + c23e_{3z} \tag{6.46}$$

$$q_5 = \arctan\left\{\frac{-s4(c1e_{3x} + s1e_{3y}) - c4[c23(-s1e_{3x} + c1e_{3y}) + s23e_{3z}]}{-s23(-s1e_{3x} + c1e_{3y}) + c23e_{3z}}\right\} \tag{6.47}$$

For determination of q_6 we equate the 1,1 and 1,2 elements in Equations 6.5 and 5.34 and then we get the sine and cosine of q_6

$$c6 = c4(c1e_{1x} + s1e_{1y}) + s4[c23(-s1e_{1x} + c1e_{1y}) + s23e_{1z}] \tag{6.48}$$

$$s6 = -c4(c1e_{2x} + s1e_{2y}) - s4[c23(-s1e_{2x} + c1e_{2y}) + s23e_{2z}] \tag{6.49}$$

$$q_6 = \arctan\left\{\frac{-c4(c1e_{2x} + s1e_{2y}) - s4[(c23(-s1e_{2x} + c1e_{2y}) + s23e_{2z}]}{c4(c1e_{1x} + s1e_{1y}) + s4[c23(-s1e_{1x} + c1e_{1y}) + s23e_{1z}]}\right\}$$

$$\tag{6.50}$$

The method of the inverse task solution demonstrated above for the Unimation PUMA 600 manipulator is also valid for kinematically simple manipulators. Most of the commercially available manipulators can be dealt with by the application of the method presented. There are some pitfalls in solving the inverse task but fortunately there are ways to avoid them. These are, for instance:

(1) Division by the sine or cosine of an angle, e.g. after having determined

q_5, we could get s6 and c6 by dividing the 2,1 and 2,2 elements of the left-hand side of Equation 6.5 by c5. This method, however, would lead to inaccuracy when c5 is near or equal to zero.

(2) The reduction of the use of common terms, wherever possible to achieve it by carrying out the premultiplication by the inverse transformation matrices of already known joint angles.

6.4 DECOMPOSITION TECHNIQUE FOR THE SOLUTION OF THE INVERSE TASK

In this section we present another relatively simple solution to the inverse task for manipulators with six degrees of freedom and in which the last three rotational axes q_4, q_5 and q_6 intersect in one point in the wrist. This allows the decomposition of the problem into two simpler ones [6.4].

The property of three intersecting joint axes allows us to decompose the task of six unknown joint variables into two tasks of three unknowns each. As a consequence of this convenient geometrical arrangement we can first determine the position and orientation of the wrist with respect either to the manipulator's TCP or to the wrist flange. Then, by already knowing the spatial position of the wrist, we can directly calculate the first three joint variables q_1, q_2 and q_3, since only these variables can influence the position of the wrist joints. After having independently calculated q_1, q_2 and q_3, we can use them to calculate the other three joint angles, q_4, q_5 and q_6. The calculation of these angles will be similar to that of Euler angles from the orientation matrix.

The solution to be outlined here combines a straightforward geometrical method with the systematic use of joint transformation matrices, thus leading to a solution using a minimum amount of arithmetic.

In the following we present the method of calculating the wrist position of the symmetrically structured Hitachi assembly manipulator. Figure 6.1 shows that the orientation of the hand described by the Euler angles α_h, β_h and γ_h and its position given by the coordinates (x_h, y_h, z_h) directly determine where the wrist must be placed in the workspace. By using the notation of Figure 6.1 the wrist position is described by

$$x_w = x_h - D \cos \alpha_h \sin \beta_h \tag{6.51}$$

$$y_w = y_h - D \sin \alpha_h \sin \beta_h \tag{6.52}$$

$$z_w = z_h - D \cos \beta_h \tag{6.53}$$

where D is the distance between the wrist intersection point and the TCP. We note that four multiplications are needed to calculate the wrist position.

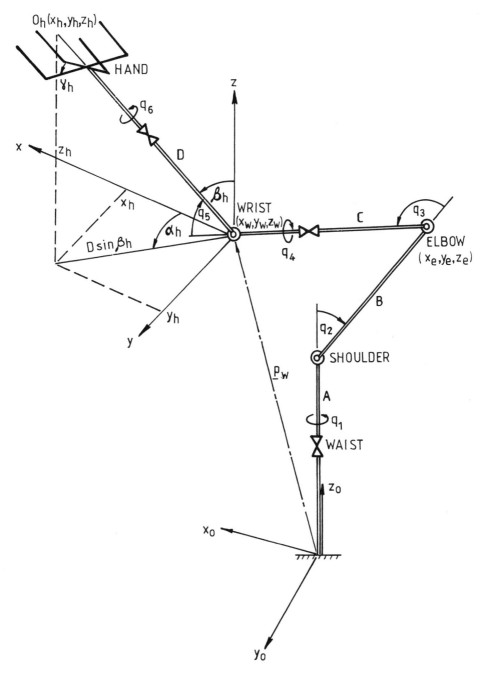

Figure 6.1 Graphical demonstration of calculating the wrist positions.

To calculate (q_1, q_2, q_3) we express the wrist position as the result of the joint transformations described in Section 5.2, i.e. we specify how the (q_1, q_2, q_3) joint rotations carry the wrist in the position defined by the coordinates (x_w, y_w, z_w). The matrix $H_{0,3}$ in Equation 5.29b specifies the transformation from the base to the elbow

$$
H_{0,3} = \begin{bmatrix}
c1 & -s1c23 & s1s23 & Bs1s2 \\
s1 & c1c23 & -c1s23 & -Bc1s2 \\
0 & s23 & c23 & Bc2+A \\
0 & 0 & 0 & 1
\end{bmatrix}
\tag{6.54}
$$

The position of the wrist in the coordinate system (x_e, y_e, z_e) fixed to the elbow is $(0, 0, C)$ where C is the length of the link between the elbow and the wrist. We transform this point into the coordinate system of the base by the homogeneous transformation matrix to elbow $H_e = H_{0,3}$, namely

$$
\begin{bmatrix} x_w \\ y_w \\ z_w \\ 1 \end{bmatrix} = H_{0,3} \begin{bmatrix} 0 \\ 0 \\ C \\ 1 \end{bmatrix}
\tag{6.55}
$$

where

$$
x_w = s1(Cs23 + Bs2)
\tag{6.56}
$$

$$
y_w = -c1(Cs23 + Bs2)
\tag{6.57}
$$

$$
z_w = Cc23 + Bc2 + A
\tag{6.58}
$$

where

$$
s23 = \sin (q_2 + q_3)
$$
$$
c23 = \cos (q_2 + q_3)
$$

and A, B, C are the subsequent link lengths.
    In Equations 6.56 and 6.57 let

$$
K = Cs23 + Bs2
\tag{6.59}
$$

then

$$x_w = Ks1 \tag{6.60}$$

$$y_w = -Kc1 \tag{6.61}$$

from which

$$s1 = \frac{x_w}{K} \qquad c1 = -\frac{y_w}{K} \tag{6.62}$$

and

$$q_1 = \arctan \frac{-x_w}{y_w} \tag{6.63}$$

From Equation 6.62 we obtain

$$K^2 = x_w^2 + y_w^2 \tag{6.64}$$

$$K = \pm\sqrt{x_w^2 + y_w^2} \tag{6.65}$$

In Equation 6.58 let

$$L = Cc23 + Bc2 \tag{6.66}$$

then from Equation 6.58

$$L = z_w - A \tag{6.67}$$

thus K and L become known quantities.

From Equations 6.59 and 6.66 we get

$$K^2 = C^2 s23^2 + B^2 s2^2 + 2BC s23 s2 \tag{6.68}$$

$$L^2 = C^2 c23^2 + B^2 c2^2 + 2BC c23 c2 \tag{6.69}$$

and by adding Equation 6.68 to Equation 6.69, we obtain

$$K^2 + L^2 = B^2 + C^2 + 2BC(c23c2 + s23s2) \tag{6.70}$$

where

$$c23c2 + s23s2 = \cos[(q_2 + q_3) - q_2] = \cos q_3$$

i.e.

$$c3 = \frac{K^2 + L^2 - (B^2 + C^2)}{2BC} \tag{6.71}$$

$$s3 = \pm \sqrt{1 - \left(\frac{K^2 + L^2 - (B^2 + C^2)}{2BC}\right)^2} \tag{6.72}$$

and finally

$$q_3 = \arctan\left[\frac{\pm\sqrt{4B^2C^2 + (B^2 + C^2)^2 - (K^2 + L^2)^2}}{K^2 + L^2 - (B^2 + C^2)}\right\} \tag{6.73}$$

In Equation 6.73 the positive and negative sign signifies whether the elbow is in an up or down configuration.

To calculate q_2 we consider Equations 6.59 and 6.66 again

$$K = Cs2c3 + Cc2s3 + Bs2 \tag{6.74}$$

$$L = Cc2c3 - Cs2s3 + Bc2 \tag{6.75}$$

Let us now introduce

$$M = Cc3 \quad \text{and} \quad N = Cs3 \tag{6.76}$$

where M and N are now known quantities, and then from Equations 6.74 and 6.75 we obtain

$$K = (M + B)s2 + Nc2 \tag{6.77}$$

$$L = (M + B)c2 - Ns2 \tag{6.78}$$

and from the values of K and L both $\sin q_2$ and $\cos q_2$ are resolved as

$$s2 = \frac{K(M + B) - LN}{(M + B)^2 + N^2} \tag{6.79}$$

$$c2 = \frac{L(M + B) + KN}{(M + B)^2 + N^2} \tag{6.80}$$

$$q_2 = \arctan\frac{K(M + B) - LN}{L(M + B) + KN} \tag{6.81}$$

The geometrical meaning of the symbols A, B, C, K, L, M and N introduced in the calculation of the q_1, q_2 and q_3 angles is shown in Figure 6.2. It is obvious that the workspace of the manipulator is limited and that a solution does not exist for q_1, q_2 and q_3 for every wrist position.

Another problem to be considered is that the inverse trigonometric functions used to calculate the joint angles have multiple solutions and we therefore have to choose carefully the right option. In solving this problem we analyze the workspace of the manipulator in the next section.

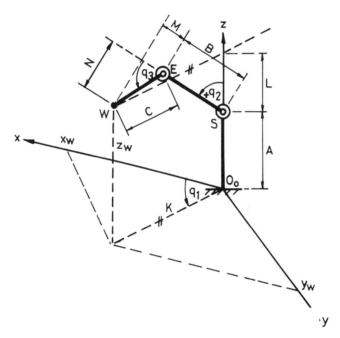

Figure 6.2 Calculating q_1, q_2 and q_3.

6.5 MANIPULATOR'S WORKSPACE DESIGN

In this section four aspects of robot workspace design will be outlined.

(1) By knowing the wrist position values of x_w, y_w, z_w derived in Section 6.4, we can check whether the wrist can be carried into the desired position or not. The first three links of the robot (A, B, C) are in the same plane. From Figure 6.3 we can see the areas of the workspace which cannot be reached by the wrist. The following observations can be made:

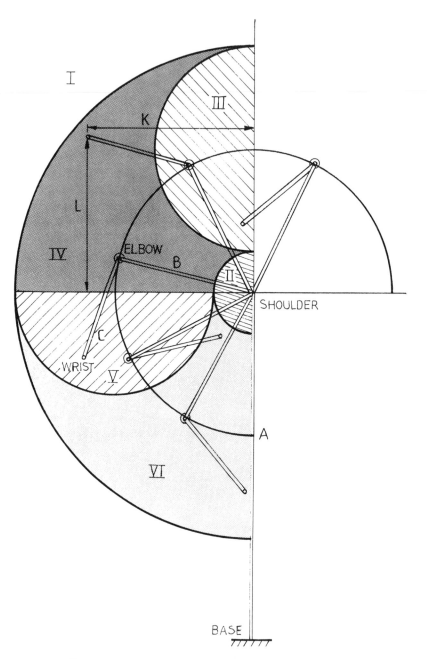

Figure 6.3 The workspace of an ideal manipulator.

(a) For those points which are too far from the shoulder, the following inequality is valid

$$K^2+L^2>(B+C)^2 \qquad (6.82)$$

(area I)

(b) For those points which are too close to the shoulder, the following inequality is valid

$$K^2+L^2<(B-C)^2 \qquad (6.83)$$

(area II)

(c) When the robot links are of equal length and the joint at the elbow has the full range of motion $\pm 180°$ the problem above does not arise.

(2) When calculating q_1 we used the arctan function. If $y_w<0$ in Equation 6.63 then the range $[0°, 180°]$ given by the arctan function is the correct solution. However, if $y_w>0$ then the correct solution is

$$q_1=\arctan\left(\frac{-x_w}{y_w}\right)-180° \qquad (6.84)$$

Thus the entire solution range for q_1 is $[-180°, 180°]$.

Equation 6.73 leads to both a positive and a negative value for q_3. This means that the same wrist position can be produced by different joint-link configurations. If there are no other conditions, such as continuous joint trajectories, or avoiding obstacles or an explicit command from the operator to change configuration, then we may choose q_3 as

$$q_3\geq 0 \qquad (6.85)$$

This expresses the usual configuration that the elbow is over the shoulder-wrist line as is shown in Figure 6.3.

(3) Calculating the values of K: since the angles (q_1, q_2, q_3) and $(180°+q_1, -q_2, -q_3)$ produce the same wrist position we can fix it as

$$K=\sqrt{x_w^2+y_w^2}\geq 0 \qquad (6.86)$$

i.e. we shall consider only the left half of the coordinate system fixed to the shoulder:

$$x_w^{\text{shoulder}}\geq 0 \qquad (6.87)$$

If in the coordinate system of the shoulder, wrist positions occur only in this area of the workspace then

$$\psi = q_2 + q_3 > 0 \qquad\qquad (6.88)$$

will always hold true, and is demonstrated for an ideal manipulator in Figure 6.4.

(4) The range of q_2: in Equation 6.81 which expresses the calculation of q_2 for which we have

$$K \geq 0$$

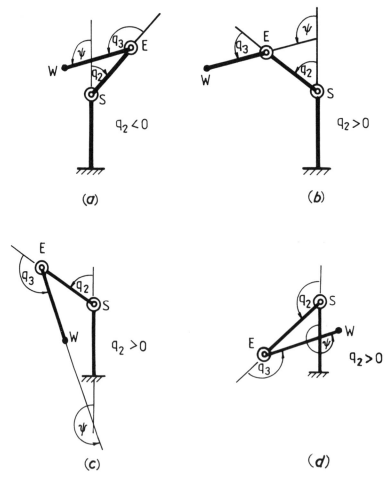

Figure 6.4 The operating range of $\psi = (q_2 + q_3)$. (*a*) Area III. (*b*) Area IV. (*c*) Area V. (*d*) Area VI.

and

$$L \geq 0 \text{ if } z_w \geq A \quad \text{and} \quad L < 0 \text{ if } z_w < A \tag{6.89}$$

The arctan function gives the range $[-90°, 90°]$ as a result for q_2. The other solution would be $q_2' = 180° - q_2$. By considering the operating range in Figure 6.5, we can decide which solution should be chosen.

(a) If

$$L > 0 \quad \text{and} \quad T < C \tag{6.90}$$

then

$$q_2 < 0$$

where

$$T = \sqrt{(B-L)^2 + K^2} \tag{6.91}$$
$$\text{(area III)}$$

Wrist positions in area III in Figure 6.3 can be reached only if the elbow is to the left of the base–shoulder line, provided that the condition in Equation 6.85 for q_3 holds (see Figure 6.5). Since $q_2 \geq -90°$ is always true there is no need to check the conditions in Equation 6.90, the arctan subroutine will always give the correct value for q_2.

(b) In area IV in Figure 6.3 the value of q_2 obtained by the arctan function can always be accepted, since here $0 \leq q_2 \leq 90°$ and

$$L > 0, \ T > C, \ U > C \tag{6.92}$$

where

$$U = [(K-B)^2 + L^2]^{1/2} \tag{6.93}$$
$$\text{(area IV)}$$

(c) If the wrist is in the area V, then the elbow–wrist line intersects the P–shoulder line in Figure 6.5 (b) and the following inequality is valid

$$0° \leq q_2 \leq 90°$$

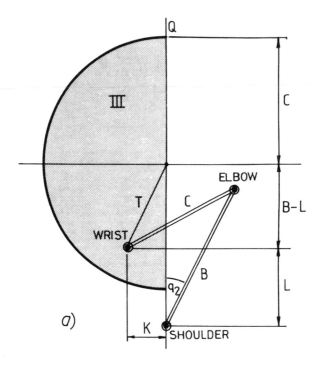

a)

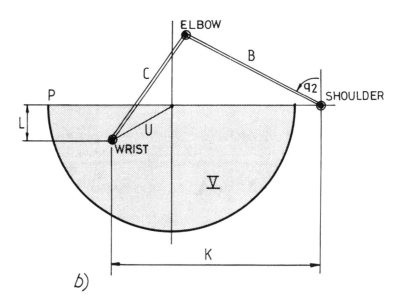

b)

Figure 6.5 The operating range of q_2. (*a*) Area III. (*b*) Area V.

and its value is given by the arctan function where

$$L < 0 \quad \text{and} \quad U < C \tag{6.94}$$

(area V)

(d) Finally in area VI, $90° < q_2 \leq 180°$ will be true, that is

$$q_2 = 180° - \arctan\left[\frac{K(M+B)-LN}{L(M+B)+KN}\right] \tag{6.95}$$

This will be the correct solution instead of Equation 6.81 if

$$L < 0 \quad \text{and} \quad U > C \tag{6.96}$$

(area VI)

The conditions outlined above need to be checked by the robot control program.

6.6 FURTHER ASPECTS OF ROBOT DESIGN

(1) Conditions of choice of robot arm length

(a) As we mentioned in the case of ideal bending joints it is not useful to make $B \neq C$, since the inside of the sphere of radius $(B-C)$ cannot be reached by the wrist.

(b) Since the range of bending joints is usually less than the ideal $\pm 180°$, condition (a) is not necessarily the case. $C > B$ must be excluded, however, in any case the elbow–wrist link must not be longer than the shoulder–elbow link. This would not only lead to an unreachable spherical range but the whole mechanical stability would deteriorate when compared to the case of $C < B$, and in addition would cause unnecessary wear and tear of the arm.

(c) If the wrist has to reach points inside a circle of radius R on the ground, the manipulator has to be designed in such a way that the following relationship should be satisfied (see Figure 6.6)

$$B + C = \pm\sqrt{\delta^2 + A^2} \tag{6.97}$$

(d) The distance A from base to shoulder depends on the desired size of the entire workspace.

(e) Considering the case when the operating range of bending joints is $\pm 90°$,

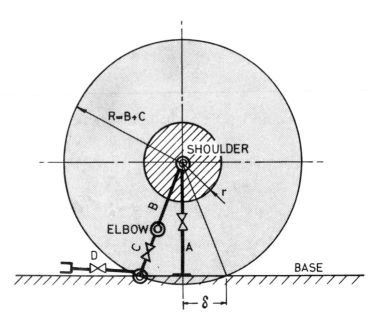

Figure 6.6   Method of choosing link lengths of the robot.

only the points outside the sphere of radius B around the shoulder and in area III, area IV or area V in Figure 6.3 can be reached by the wrist, i.e. we can control the position and the orientation of the end effector only inside these areas.

(2) Calculating q_4, q_5, q_6

(a) By knowing q_1, q_2 and q_3 (see Section 6.3) we can obtain the orientation of the first three joints by calculating the orientation submatrix of the base-to-elbow transformation, i.e.

$$R_{0,3} = \begin{bmatrix} c1 & -s1c23 & s1s23 \\ s1 & c1c23 & -c1s23 \\ 0 & s23 & c23 \end{bmatrix} \qquad (6.98)$$

This is the same as the upper left 3×3 submatrix of the transformation matrix in Equation 5.29b.

(b) The desired orientation of the hand in the coordinate system of the base is given in terms of the orientation matrix in Equation 3.5 composed of the Euler angles. We denote this matrix by R_E as shown in Figure 6.7.

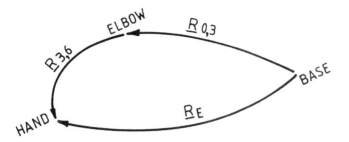

Figure 6.7   Relationship of orientation matrices.

(c) The base-to-hand orientation matrix can be decomposed as

$$R_E = R_{0,3} * R_{3,6} \tag{6.99}$$

and thus

$$R_{3,6} = R_{0,3}^{-1} * R_E \tag{6.100}$$

Since orientation matrices are orthogonal, the relation for the inverse of matrix $R_{0,3}$ is

$$R_{0,3}^{-1} = R_{0,3}^T \tag{6.101}$$

thus

$$R_{3,6} = R_{0,3}^T * R_E \tag{6.102}$$

(d) On the other hand the elements of the rotation partition of $H_{3,6}$ denoted by $r_{i,j}$ where $i, j = 1, 2, 3$ are

$$R_{3,6} = \begin{bmatrix} c4c6 - s4c5s6 & -c4s6 - s4c5c6 & s4s5 \\ s4c6 + c4c5s6 & -s4s6 + c4c5c6 & -c4s5 \\ s5s6 & s5c6 & c5 \end{bmatrix}$$

$$= \begin{bmatrix} r_{11} & r_{12} & r_{13} \\ r_{21} & r_{22} & r_{23} \\ r_{31} & r_{32} & r_{33} \end{bmatrix} \tag{6.103}$$

The angles q_4, q_5 and q_6 correspond to the angles α, β and γ, i.e. the matrix in Equation 6.103 is the same as that of Equation 3.5. For this reason we can use the same algorithm to calculate them as we have used for the calculation of the Euler angles from the orientation matrix in Sections 3.2, 3.4 and 3.6. This also explains the particular choice of the wrist geometry configuration in terms of Euler angles. They correspond to the given joint arrangement of the manipulator in question. Thus we also include the choice $0° < q_5 < 180°$ which is the last of three options which makes the manipulator configuration unique for a given hand position and orientation.

(3) Further aspects on choice of arm geometry configuration

(a) The range of joint angles chosen can be used without any further consideration of their values if the manipulator has to be guided from one state (e.g. from a parking position) to another state (e.g. to the starting point) without any regard to the locus of the joints while they move. During motion along a trajectory, another point of primary importance is that the joint angles change continuously, and a sudden change of joint angles between the two end points of a trajectory segment must not be allowed to occur.

(b) It may happen that the initial conditions imposed on joint angle ranges do not correspond to the real situation, e.g. since the elbow or the wrist could collide with the workspace environment. We can often avoid such collisions by choosing a new configuration, while leaving the hand path unchanged. Taking into account the above aspects, the range of joint angles must be chosen as follows:

   (i) If a higher level robot control program specifies a definite setting of joints along the trajectory, then the control program sets these ranges while positioning the end effector at the start point of the segment.
   (ii) If there is no definite setting then the ranges mentioned in the previous sections are set.
   (iii) While continuously traveling along a trajectory segment, the program always has to choose joint angles in such a way that no sudden change could occur, i.e. it tries to keep within the already existing ranges.
   (iv) If for some reason the manipulator configuration must be changed, e.g. because of running out of range of any of the joint angles, then this may happen only by slicing up the trajectory.

(4) Summary of computation

In the listing of operations for computation in Table 6.1 we disregard all additions and subtractions because they require considerably less time than any other operation. But we consider computation of all the multiplications,

Table 6.1

Status or action	Number of operations*
Wrist position	4 multiplications
	(Eqs. 6.51–6.53)
Calculating q_1	1 division (Eq. 6.63)
Calculating K^2	2 multiplications (Eq. 6.64)
Calculating K	1 square root (Eq. 6.65)
Calculating L^2	1 multiplication
Calculating q_3	1 division (Eq. 6.73)
	1 square root (Eq. 6.73)
Calculating q_2	10 multiplications (Eq. 6.81)
	1 division (Eq. 6.81)
Developing the $R_{0,3}$ matrix	8 multiplications (Eq. 6.98)
Developing the R_E matrix	12 multiplications (Eq. 3.5)
Developing the $R_{3,6}$ matrix	27 multiplications (Eq. 6.102)
Calculating q_5	4 multiplications
	1 division
	1 square root (Eq. 3.28)
Calculating q_4	4 multiplications
	1 division (Eq. 3.32)
Calculating q_6	4 multiplications
	1 division (Eq. 3.36)

*Additions and subtractions are not listed since their time demand is negligible in comparison with other arithmetic operations.

divisions, squares, square roots and the trigonometrical functions. Trigonometrical functions can be evaluated from look-up tables by using interpolation. By this method considerable processing time may be saved.

The constant values B^2, C^2, $2BC$, $4B^2C^2$, $(B+C)^2$, $(B-C)^2$ can be calculated in advance and stored in memory. The square root calculation corresponds to about ten divisions when using integer arithmetic. Taking all this into consideration the six joint angles can be determined by 112 arithmetic operations.

6.7   APPENDICES

Appendix 6.A Evaluation of matrix $[a_1\ a_2\ a_3\ a_k]$ in Equation 6.26

$$H_{1,2}^{-1} * H_{0,1}^{-1} * T_6 = [a_1\ a_2\ a_3\ a_k] \tag{6.26}$$

The column vectors on the right-hand side are expressed in expanded form as

$$a_1 = \begin{bmatrix} a_{1X} \\ a_{1Y} \\ a_{1Z} \\ 0 \end{bmatrix} = \begin{bmatrix} c1e_{1x} + s1e_{1y} \\ c2(-s1e_{1x} + c1e_{1y}) + s2e_{1z} \\ -s2(-s1e_{1x} + c1e_{1y}) + c2e_{1z} \\ 0 \end{bmatrix} \tag{6.A.1}$$

$$a_2 = \begin{bmatrix} a_{2X} \\ a_{2Y} \\ a_{2Z} \\ 0 \end{bmatrix} = \begin{bmatrix} c1e_{2x} + s1e_{2y} \\ c2(-s1e_{2x} + c1e_{2y}) + s2e_{2z} \\ -s2(-s1e_{2x} + c1e_{2y}) + c2e_{2z} \\ 0 \end{bmatrix} \tag{6.A.2}$$

$$a_3 = \begin{bmatrix} a_{3X} \\ a_{3Y} \\ a_{3Z} \\ 0 \end{bmatrix} = \begin{bmatrix} c1e_{3x} + s1e_{3y} \\ c2(-s1e_{3x} + s1e_{3y}) + s2e_{3z} \\ -s2(-s1e_{3x} + c1e_{3y}) + c2e_{3z} \\ 0 \end{bmatrix} \tag{6.A.3}$$

$$a_k = \begin{bmatrix} a_{kX} \\ a_{kY} \\ a_{kZ} \\ 1 \end{bmatrix} = \begin{bmatrix} c1k_x + s1k_y \\ c2(-s1k_x + c1k_y) + s2k_z \\ -s2(-s1k_x + c1k_y) + c2k_z \\ 1 \end{bmatrix} \tag{6.A.4}$$

Appendix 6.B Evaluation of matrix $[b_1\ b_2\ b_3\ b_k]$ in Equation 6.29

$$H_{2,3}^{-1} * H_{1,2}^{-1} * H_{0,1}^{-1} * T_6 = [b_1\ b_2\ b_3\ b_k] \tag{6.29}$$

The column vectors on the right-hand side are expressed in expanded form as

$$b_1 = \begin{bmatrix} b_{1X} \\ b_{1Y} \\ b_{1Z} \\ 0 \end{bmatrix} = \begin{bmatrix} c1e_{1x} + s1e_{1y} \\ c23(-s1e_{1x} + c1e_{1y}) + s23e_{1z} \\ -s23(-s1e_{1x} + c1e_{1y}) + c23e_{1z} \\ 0 \end{bmatrix} \tag{6.B.1}$$

$$b_2 = \begin{bmatrix} b_{2X} \\ b_{2Y} \\ b_{2Z} \\ 0 \end{bmatrix} = \begin{bmatrix} c1e_{2x} + s1e_{2y} \\ c23(-s1e_{2x} + c1e_{2y}) + s23e_{2z} \\ -s23(-s1e_{2x} + c1e_{2y}) + c23e_{2z} \\ 0 \end{bmatrix} \tag{6.B.2}$$

$$b_3 = \begin{bmatrix} b_{3X} \\ b_{3Y} \\ b_{3Z} \\ 0 \end{bmatrix} = \begin{bmatrix} c1e_{3x} + s1e_{3y} \\ c23(-s1e_{3x} + c1e_{3y}) + s23e_{3z} \\ -s23(-s1e_{3x} + c1e_{3y}) + c23e_{3z} \\ 0 \end{bmatrix} \tag{6.B.3}$$

$$b_k = \begin{bmatrix} b_{kX} \\ b_{kY} \\ b_{kZ} \\ 1 \end{bmatrix} = \begin{bmatrix} c1k_x + s1k_y \\ c23(-s1k_x + c1k_y) + s23k_z \\ -s23(-s1k_x + c1k_y) + c23k_z \\ 1 \end{bmatrix} \tag{6.B.4}$$

where

$$s23 = \sin(q_2 + q_3)$$
$$c23 = \cos(q_2 + q_3)$$

Appendix 6.C Evaluation of matrix $[c_1 \; c_2 \; c_3 \; c_k]$ in Equation 6.43

$$H_{3,4}^{-1} * H_{2,3}^{-1} * H_{1,2}^{-1} * H_{0,1}^{-1} * T_6 = [c_1 \; c_2 \; c_3 \; c_k] \tag{6.43}$$

The column vectors on the right hand side are expressed in expanded form as

$$c_1 = \begin{bmatrix} c_{1X} \\ c_{1Y} \\ c_{1Z} \\ 0 \end{bmatrix} = \begin{bmatrix} c4(c1e_{1x} + s1e_{1y}) + s4c23(-s1e_{1x} + c1e_{1y}) + s4s23e_{1z} \\ -s4(c1e_{1x} + s1e_{1y}) + c4c23(-s1e_{1x} + c1e_{1y}) + c4s23e_{1z} \\ -s23(-s1e_{1x} + c1e_{1y}) + c23e_{1z} \\ 0 \end{bmatrix} \tag{6.C.1}$$

$$c_2 = \begin{bmatrix} c_{2X} \\ c_{2Y} \\ c_{2Z} \\ 0 \end{bmatrix} = \begin{bmatrix} c4(c1e_{2x} + s1e_{2y}) + s4c23(-s1e_{2x} + c1e_{2y}) + s4s23e_{2z} \\ c4(c1e_{2x} + s1e_{2y}) + c4c23(-s1e_{2x} + c1e_{2y}) + c4s23e_{2z} \\ -s23(-s1e_{2x} + c1e_{2y}) + c23e_{2z} \\ 0 \end{bmatrix} \tag{6.C.2}$$

$$c_3 = \begin{bmatrix} c_{3X} \\ c_{3Y} \\ c_{3Z} \\ 0 \end{bmatrix} = \begin{bmatrix} c4(c1e_{3x}+s1e_{3y})+s4c23(-s1e_{3x}+c1e_{3y})+s4s23e_{3z} \\ s4(c1e_{3x}+s1e_{3y})+c4c23(-s1e_{3x}+c1e_{3y})+c4s23e_{3z} \\ -s23(-s1e_{3z}+c1e_{3y})+c23e_{3z} \\ 0 \end{bmatrix}$$

$$(6.C.3)$$

$$c_k = \begin{bmatrix} c_{kX} \\ c_{kY} \\ c_{kZ} \\ 1 \end{bmatrix} = \begin{bmatrix} c4(c1k_x+s1k_y)+s4c23(-s1k_x+c1k_y)+s4s23k_z \\ -s4(c1k_x+s1k_y)+c4c23(-s1k_x+c1k_y)+c4s23k_z \\ -s23(-s1e_{3z}+c1e_{3y})+c23e_{3z} \\ 1 \end{bmatrix}$$

$$(6.C.4)$$

6.8   REFERENCES

[6.1] Paul, R., Shimano, B. and Mayer, G. E. Kinematic control equations for simple manipulators, *IEEE Tr. on Systems, Man and Cybernetics*, Vol. SMC-11, No. 6, June 1981.

[6.2] Albus, J. S. A new approach to manipulator control: The cerebellar model articulation controller (CMAC), *Trans. ASME*, Sept. 1975.

[6.3] Paul, R. P., *Robot Manipulators: Mathematics, Programming and Control*, The MIT Press, 1981.

[6.4] Horn, B. K. P. and Inoue, H. Kinematics of the MIT-AI VICARM manipulator, *Working Paper 69*, Massachusetts Inst. of Technology, Artificial Intelligence Lab., May 1974.

7

SERVO-SYSTEMS FOR ROBOT CONTROL

7.1   GENERAL ASPECTS OF ROBOT CONTROL

Robot control is understood to be the application of microprocessor or multi-microprocessor based control systems, which are in fact programmable microcontrollers, to carry out mechanical manipulation. The objective of the controllers is to provide all necessary information for accurate positioning with specified velocity and acceleration. A further aim is to generate the necessary torques and/or forces to minimize various types of errors inherent in the robot operation and to reduce the effects of the errors generated by external disturbances to a specified minimum level. All these may be done by a centralized control system which means that the coordination of all joint actions is controlled by a single computer based controller. On the other hand a second option is the use of a distributed control system. This is composed of a number of local microprocessor based motor controllers, each of which involves the control of a single joint-link segment with its own single variable. But in addition there is a *primus inter pares* processing computer acting in a supervisory role for providing the control for functional motion. The former is usually adopted for simple pick-and-place routine work whilst the latter is for more complex robot operation for assembly work.

In a more general interpretation, the aim of robot control is to obtain a satisfactory performance of the manipulator, with specified figures of merit, regardless of which kind of servo-system organization we are dealing with. One of the most important performance factors of the robot is the number of executed operations per unit time, or alternatively its cycle time. This is in a direct relationship with the operating speed of the robot manipulator and can only be improved by increasing the speed, which requires further enhancement of the robot kinematic and dynamic performance.

From a structural point of view, the robot is a multilink mechanical transmission chain and, as such, it may be considered as a series of weakly coupled second order systems. The coupling effects are found to be weak, because even in modern robots the speeds are still relatively low. Nevertheless, the coupling effects are there and their reduction to a minimum level is of paramount importance. This can be achieved by the appropriate choice of structural inertias and by ensuring that the servo frequency bandwidth, along the joint-link segments, is appropriately set by design to exponentially increase from the shoulder joint towards the wrist as shown in the case of the human arm (see Section 1.3).

Currently available industrial robots are driven by hardware and software controlled servo-systems. The power drives may be hydraulic, pneumatic or electrical. The electronics parts of the controllers are traditionally hybrid, being analog/digital or lately completely digital and suitable for controlling second order systems with constant inertia, compliance and viscous damping. The robot systems are neither linear nor time invariant. They are in fact time-variant, nonlinear, joint interdependent devices, and as such they might cause stability and controllability problems.

A reduced interdependence of the active arm segments may be obtained by the use of sufficient feedforward compensation. For the same purpose it is normal practice in robot design to employ heavily overrated power drives in order to ease the effects of the various constraints inherently present in robot operation. Under such conditions the system parameters remain more stationary and do not vary so abruptly with the variation of the robot's kinematic geometry. These are only a few of the factors worth mentioning in order to understand how the robot joint-link arm segments can be made nearly independent. In doing so we will show how a set of second order time-invariant linear systems can be adapted to model the robot in low speed operation and why we may succeed by using currently available linear controllers.

7.2 BASIC CONTROL TECHNIQUES

In this section we are concerned with the interpretation of servo operation in robot manipulator systems. We shall now carry out a study of robot performance for position, velocity and acceleration. To start with we shall recall some important servo control principles and revise the relevant theories necessary to demonstrate the special methods required for robot modelling shown in Section 7.3.

For the description of robot servo operation we will apply classical servo principles [7.1], [7.2] and derive the responses in both time and frequency domains. Robot engineering embraces servo-system theory and considerably extends it. One of its principal features is to control the output quantities from a single input or multiple inputs. The essential schematics of a basic automation system are shown in Figure 7.1.

The robot servo operates in a closed loop and consists of three essential parts: an intelligence unit for comparison, a forward path for power drive control, and often more than one feedback loop connected to position and velocity transducers. In addition a special feature for robots incorporates more than one feedforward loop for various steady-state and random error compensation. The forward path unit represents the power, i.e. it is the muscle of the robot arm segment. The comparison unit, the feedback and feedforward loops provide the quality properties of the system, i.e. the hardware intelligence or autonomous type features like jerk reflex reaction, known in humans as efferent action via the ganglia centers. These are the stationary, inflexible intelligence properties. The flexible intelligence resides in the software, i.e. in the user and system programs located in RAMs and PROMs respectively.

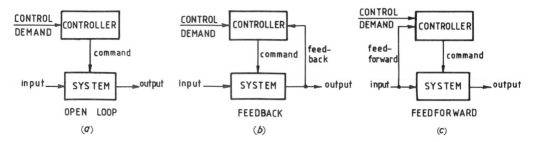

Figure 7.1 Three essential schematics of an automation system. (*a*) Command automation for creating the output without sensing the process (system). (*b*) Command–feedback automation for creating the output, using information sensed from the output of the process in addition to the control command. (*c*) Command–feed-forward automation for creating the output, using information sensed from the input of the process in addition to the control command.

In servo-system operation the comparison unit compares the actual output with the reference input. Any difference results in an error which is amplified and fed forward to control the flow of power to the load. The output is the load's condition in terms of position, orientation, velocity and acceleration. These may be fed back directly or via the modifying feedback loops to the comparison unit. Since it is a servo that is error actuated, there is the facility to compensate for any departure of the output from the desired conditions set by the input. It also compensates for any unwanted variation in the properties of the forward paths and feedback loops like aging, wear and tear, etc. There are also external and internal disturbances which may be encountered anywhere in the system and may generate errors, which also have to be dealt with.

The basic servo principles in the following sections will apply to any control system whatever its physical structure. It may be electrical, electrome-chanical, hydraulic, pneumatic or any combination of these, but the electronic and electrical techniques have many advantages. They are very convenient for signal comparison, error detection, amplification and data transmission. The amplified error signal, i.e. the control strategy, will then be used to control directly the electrical or indirectly the hydraulic or pneumatic servo actuators.

7.3   MATHEMATICAL MODELING OF ROBOT SERVOS

In robot operation the primary concern is smooth motion and error-free positioning. The main objective now is to derive the robot joint-link's servo responses in order to quantify the position, velocity and acceleration errors and then to devise compensation techniques for the improvement of robot kinematic performance. In order to achieve these the following steps are suggested:

(i) Derivation of a position servo model of the robot joint-link segment in both time and frequency domains.
(ii) Provision of the transfer function and a block diagram representation.
(iii) Modeling of servo responses irrespective of load and position variations.

(i) From a control point of view, the robot arm manipulator at low speed can be considered as a series of weakly coupled servo-systems, each servo controlling a single joint-link segment. The number of servos required depends upon the number of joints. Each servo-system is characterized by a variable inertia or mass, and a mechanical stiffness resulting in a structural resonant frequency. It has already been mentioned in Section 5.3 that, from a mathematical point of view, there is no difference between rotational connections with transverse or longitudinal axes, i.e. between bend and swivel joints. Such normalization of the joints greatly simplifies the modeling of robot systems. In the following discussion we shall deal with the robot as a mechanical chain, decomposed into independent joint-link segments, on which both the first- and second-order effects can easily be dealt with. The joint-link segmented configurations of a six-axis robot manipulator are shown in Figure 7.2.

The investigation starts at the last joint J6 adjacent to the end effector which is left free to move as shown in Figure 7.2(*a*). The rest of the joints encircled are rigidly fixed. We will then proceed from joint to joint to consider the system's effective loads. This will lead to a simplified, but very general treatment of a single joint-link arm segment by which we can identify the major first and second order effects. The higher order effects inherently present in the robot arm segments will temporarily be disregarded, although their macroscopic effects will be considered separately later.

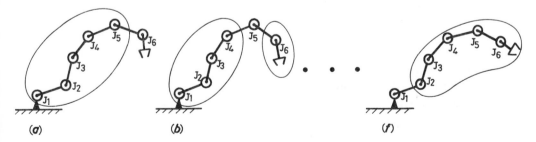

(*a*) (*b*) (*f*)

Figure 7.2 Decomposed manipulator joint-link mechanism. The encircled joints are assumed to be rigidly fixed, and the remaining one is free to provide angular displacement.

In the following treatment we will deal with an electromechanical servo-system where the mechanical drive unit consists of an armature controlled d.c. servomotor and a gear train of high reduction ratio n. The servos used for robotic drive applications must be at least Type-1 systems [7.3], as will be proved later on. A Type-1 system contains one integrator in the forward path to produce the torque integral of the motor. It is worth mentioning that the forward transfer function defines the system type and the 'type' designation is based upon the order of the exponent of the single standing Laplace operator s in the denominator.

The block diagram in Figure 7.3 represents a position and velocity servo-system which will then be extended to study also its acceleration properties. The velocity input in Figure 7.3 is also implied but at this stage there is no speed setting in the following system modeling.

The position input signal θ_p (rad) acting on the system results in an output signal θ_o (rad). The output θ_o via a position sensor is fed back and the difference between input and output is proportional to the position error signal e_p (V) given by

$$e_p = k_p(\theta_p - \theta_o) \tag{7.1}$$

where k_p is the position feedback gain coefficient; transducer voltage per unit angle (V/rad).

There is another feedback loop from the output via a tachometer device which produces a signal e_v (V) proportional to the velocity of the output shaft rotation

$$e_v = -k_v \frac{d\theta_o}{dt} \tag{7.2}$$

where k_v is the velocity feedback gain coefficient; transducer voltage per unit rate [V/(rad/sec)].

Note that we use the differential operator in the derivation of equations of motion instead of introducing Laplace transforms of the variables at this stage, but bear in mind that later on we introduce the Laplace transformation and all time-invariant variables will have Laplace transforms [7.3].

Both feedback signals e_p and e_v act in a superimposed manner as an effective error signal. This is then amplified resulting in current control strategy i for driving the robot actuators.

$$i = A\left[k_p(\theta_p - \theta_o) - k_v \frac{d\theta_o}{dt} \right] \tag{7.3}$$

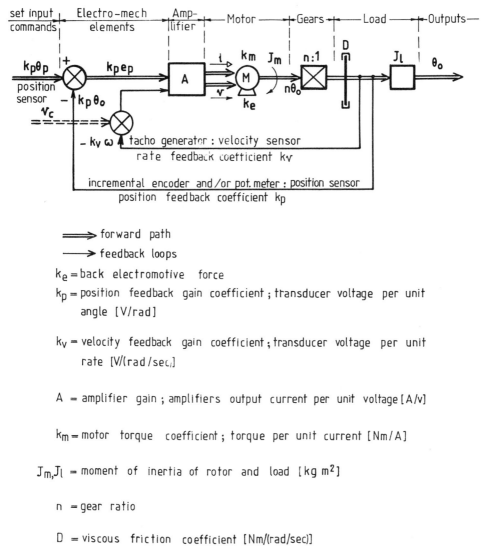

Figure 7.3 Block schematic diagram representing a joint-link servo-system actuated by an armature controlled d.c. servomotor via a gear-train.

where A is the amplifier gain, i.e. amplifier output current per unit voltage (A/V).

The gross torque $T_g = k_m i$ (N m) developed by the servomotor on the rotor shaft is expressed as

$$T_g = A k_m \left[k_p(\theta_p - \theta_o) - k_v \frac{d\theta_o}{dt} \right] \qquad (7.4)$$

where k_m is the motor torque coefficient, i.e. torque per unit current (N m/A).

Having considered how the torque is generated, we will now determine how it will be utilized to control the individual joint-link arm segments. A substantial part of the torque is absorbed in accelerating the rotor moment of inertia J_m (kg m^2) which also includes the inertia of the gear train on the rotor shaft. Consider at this stage the only load to be the moment of inertia J_m. Then the torque T_m (N m) required to accelerate this moment of inertia is given by

$$T_m = J_m \frac{d^2(n\theta_o)}{dt^2} = nJ_m \frac{d^2\theta_o}{dt^2} \tag{7.5}$$

After subtracting T_m from the gross torque T_g, the remaining torque T_r is left to control all loads on the motor shaft. Thus the useful torque is

$$T_r = T_g - T_m = T_g - nJ_m \frac{d^2\theta_o}{dt^2} \tag{7.6}$$

In a robot control system the gear-train is one of the most essential but the most problem-causing component of the mechanical structure. Its function is to transmit the power and adjust the power flow from the actuator to the load on each arm segment. The gears used in a robot drive have quite high ratios as illustrated in a comparison of two robot gear systems in Table 7.1 (see also Appendix 7.A).

Table 7.1 Gear ratios of a five degrees of freedom and a six degrees of freedom robot.

Robot type	Joint						Degrees of freedom
	J1	J2	J3	J4	J5	J6	
Syke 600–5 PUMA	160	160	192	90	90.0	–	5
Unimation	62	108	53.7	76	71.9	76.6	6

The torque T_r will act on the shaft of the motor; therefore its action must be reflected in the slow running shaft of the gear-train. The reduction gear magnifies the torque by the gear ratio n; thus the useful torque, on the external slow rotating shaft after the gear-train, can be expressed as

$$nT_r = n\left(T_g - nJ_m \frac{d^2\theta_o}{dt^2}\right) \tag{7.7}$$

The load on the slow rotating shaft consists of the inertias of the secondary part of the gear unit, the effective inertia of the arm segment, the gripper and the tool with the payload on it. Combining them we get the overall effective inertia designated by J' and reflected to the end of the arm segment. The tool itself may also act as a distributed load which depends on its position along the trajectories of motion during the robot operation. The useful torque $T' = nT_r$ is utilized in accelerating the effective overall inertia J' on the output shaft of the load, i.e.

$$T' = J' \frac{d^2\theta_o}{dt^2} B \tag{7.8}$$

In addition some part of the useful torque also has to overcome the viscous and coulomb friction, and stiction. Viscous friction is proportional to velocity, while coulomb friction is constant, and stiction is instantaneous. For the sake of simplicity we will temporarily exclude the coulomb friction and stiction from our modeling scheme and take their effects into account later on. Thus, the term for the mechanical damping torque with viscous damping factor D [N m/(rad/sec)] becomes

$$T_D = D \frac{d\theta_o}{dt} \tag{7.9}$$

The required torque for driving the effective load and the useful torque generated by the motor are equal. By equating them the differential equation describing the motion of the robot joint-link arm segment is

$$J' \frac{d^2\theta_o}{dt^2} + D \frac{d\theta_o}{dt} = n\left(T_g - nJ_m \frac{d^2\theta_o}{dt^2} \right). \tag{7.10}$$

Substituting Equation 7.4 in Equation 7.10 yields

$$J' \frac{d^2\theta_o}{dt^2} + D \frac{d\theta_o}{dt} = n\left\{ Ak_m\left[k_p(\theta_p - \theta_o) - k_v \frac{d\theta_o}{dt} \right] - nJ_m \frac{d^2\theta_o}{dt^2} \right\} \tag{7.11}$$

Rearranging Equation 7.11 in diminishing order of the differential terms and normalizing the rearranged equation, we obtain a second order linear differential equation, as the equation of motion

$$\frac{d^2\theta_o}{dt^2} = \frac{D + Ak_m nk_v}{n^2 J''} \frac{d\theta_o}{dt} + \frac{Ak_m nk_p}{n^2 J''} \theta_o = \frac{Ak_m nk_p}{n^2 J''} \theta_p \tag{7.12}$$

where

$$n^2 J'' = J' + n^2 J_m$$

Note that any variation of the effective joint-link inertia may be well masked by the constant inertias of the rotor of the motor on the fast rotating side of the gear-train, i.e. $J' \ll n^2 J_m$ and therefore the overall effective inertia reflected to the output is $n^2 J'' \cong n^2 J_m$.

(ii) Now taking Laplace transforms, with zero initial conditions, and denoting $\mathscr{L}[\theta_p]$ by $\theta_p(s)$ and $\mathscr{L}[\theta_o]$ by $\theta_o(s)$ in Equation 7.12, after manipulation the equation of motion in the complex frequency domain becomes

$$s^2 \theta_o(s) + \frac{D + Ak_m nk_v}{n^2 J''} s\theta_o(s) + \frac{Ak_m nk_p}{n^2 J''} \theta_o(s) = \frac{Ak_m nk_p}{n^2 J''} \theta_p(s) \qquad (7.13)$$

where s is the Laplace operator.

Quite often in engineering practice, control system design and testing are carried out in terms of transfer functions in the complex frequency domain utilizing block diagrams [7.4]. Equation 7.13 provides the system's block diagram shown in Figure 7.4(*a*). By moving the summing point of the major feedback loop ahead and eliminating the viscous damping feedback loop by applying block algebra rules, the reduced block diagram is obtained as shown in Figure 7.4(*b*). By definition the transfer function $G(s)$ is the ratio of Laplace transforms of the output $\theta_o(s)$ and the input $\theta_p(s)$. It is deduced either from Equation 7.13 or more conveniently from the block diagrams in Figure 7.4 as

$$G(s) = \frac{\theta_o(s)}{\theta_p(s)} = \frac{Ak_m nk_p}{s^2 n^2 J'' + (D + Ak_m nk_v)s + Ak_m nk_p} \qquad (7.14)$$

Equation 7.14 is a normalized second order system of the form

$$G(s) = \frac{w_n^2}{s^2 + 2\zeta w_n s + w_n^2} \qquad (7.15)$$

where w_n is the undamped natural frequency, ζ is the damping factor or damping ratio, ζw_n is the product term which determines the rate of decay and $w_d = w_n (1 - \zeta^2)^{1/2}$ is the damped natural frequency of oscillation, i.e. the actual servo resonant frequency of the closed-loop system.

By equating the corresponding terms in Equations 7.14 and 7.15 the characteristic terms are:

$$w_n = \sqrt{\frac{Ak_m nk_p}{n^2 J''}} \qquad (7.16)$$

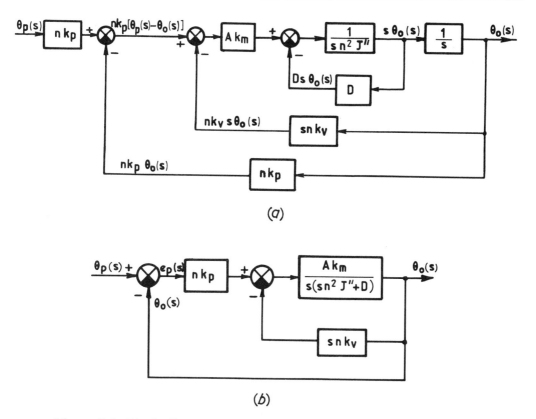

Figure 7.4 Block diagrams of a servo-controlled joint-link arm segment from Equation 7.13. (*a*) Directly decomposed version. (*b*) Simplified version.

$$\zeta w_n = \frac{D + Ak_m n k_v}{2n^2 J''} \tag{7.17}$$

$$\zeta = \frac{D + Ak_m n k_v}{2\sqrt{Ak_m n k_p n^2 J''}} \tag{7.18}$$

In general practice, servo-systems normally operate with a damping factor in the range

$$0.25 < \zeta < 0.7$$

having the corresponding per unit overshoot M_p

$$1.41 > M_p > 1.016$$

A time response for an underdamped system means that the output aligns with the input faster than it would do for a critically damped or overdamped system. It oscillates at the damped servo resonant frequency about its final value for any step, ramp or pulse input (see Figure 7.6). Such oscillation cannot be tolerated in robot operation.

(iii) The damping for robot manipulators must be critical or even overdamped. Its minimum value is $\zeta = 1$. Usually a slightly overdamped response is preferred

$$1.00 < \zeta < 1.05$$

It can easily be imagined what would happen at the boundaries of the robot workspace or with overshoots near to the bench in the case of underdamped oscillatory response. The robot manipulator would be severely damaged or, even in the slightly underdamped case, it would wear out much faster than it should, with the danger of damaging the arm by bending or twisting it out of alignment. By setting critical damping ($\zeta = 1$), Equation 7.18 becomes

$$2(Ak_m nk_p n^2 J'')^{1/2} = D + Ak_m nk_v \qquad (7.19)$$

where the amplifier gain A, gear ratio n and motor torque coefficient k_m are intended to be kept constant.

The problem regarding oscillation is that the structural natural frequencies of the joint-link arm segments vary inversely with the square root of the effective inertias between their minimum and maximum values, i.e. J''_{min} in no-load-no-stretch and J''_{max} in full-load-full-stretch positions. Both of these limit values of inertia are obtainable with relative ease by measurement.

In practice the variation of the ratio of the joint-link inertias J''_{max}/J''_{min} within the robot workspace may vary by the ratio 10:1. With inertia J''_{max} there is a corresponding minimum structural resonant frequency $w_{st,min}$ and with inertia J''_{min} a maximum structural resonant frequency $w_{st,max}$. Their ratio $w_{st,max}/w_{st,min}$ is approximately 3:1 [7.5].

From the two inertia limits above, we may derive the value of a midrange effective inertia J''_{mid} by applying a geometric and arithmetic mean averaging process until the two mean values become equal. This requirement can usually be satisfied within two steps, i.e. the averaging process is fast convergent, and then the corresponding midrange structural resonant frequency will be $w_{st,mid}$. The structural resonant frequency range for each joint-link arm segment is covered by the frequency limits as

$$w_{st,min} < w_{st,mid} < w_{st,max}$$

The midrange structural resonant frequency of each arm segment may be considered as a characteristic figure of merit of the robot arm manipulator and as a first approach we may well use this in our further mathematical treatment. The authors even suggest that the midrange structural resonant frequencies for the major joints should be given in the robot's general specification.

In order to prevent natural oscillations, there should be a minimum of 6 dB scaling down of the operating frequency spectrum below the midrange structural resonant frequency. Therefore the upper limit of the operating frequency or the servo frequency bandwidth may be chosen as $0.5 \, w_{st,mid}$, i.e.

$$\Delta w \leq 0.5 w_{st,mid} \tag{7.20a}$$

thus

$$\sqrt{\frac{A k_m n k_p}{n^2 J''_{mid}}} \leq 0.5 w_{st,mid} \tag{7.20b}$$

and

$$A k_m n k_p \leq 10 (f_{st,mid})^2 n^2 J''_{mid} \tag{7.21}$$

The inequality 7.21 serves to define the upper value of the overall servo loop gain with the characteristic midrange value of the effective inertia [7.6], i.e.

$$A k_m n k_p = 10 (f_{st,mid})^2 n^2 J''_{mid} \tag{7.22}$$

and the upper limit of the position feedback gain nk_p can be deduced from

$$nk_p \leq \frac{10 (f_{st,mid})^2 n^2 J''_{mid}}{A k_m} \approx \frac{10 f^2 n^2 J''}{A k_m} \tag{7.23}$$

Similarly the calculation can be implemented at each robot manipulator segment by setting the value of the overall loop gain $A k_m n k_p$ to calculate the corresponding position feedback gain nk_p.

It is then simple to achieve a critically damped operation by the control of the velocity feedback loop gain nk_v. All we have to do is to use the critical rate feedback coefficient $k_{v,cr}$ which is expressed by applying the midrange inertia J''_{mid}. The minimum value of nk_v for $\zeta = 1$ will be derived from Equation 7.19 in the form

$$D + k_m n k_{v,cr} = 2 (n k_p A k_m n^2 J''_{mid})^{1/2} \tag{7.24}$$

Dividing Equation 7.19 by Equation 7.24 we obtain the velocity feedback gain for any value of inertia within its maximum and minimum limits

$$nk_v \leq \left[\frac{D + Ak_m nk_{v,cr}}{\sqrt{J''_{mid}}} \sqrt{J''} - D \right] \frac{1}{Ak_m} \tag{7.25}$$

For a critically damped response the velocity feedback coefficient nk_v must be equal to the right-hand side of Equation 7.25 and the output response will be independent of the variation of effective load inertia J''. The block diagram for the critically damped operation is shown in Figure 7.5.

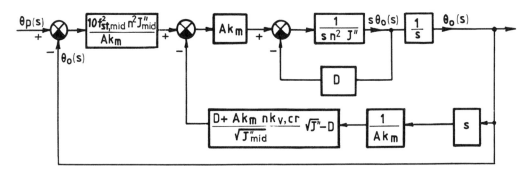

Figure 7.5 Block diagram of the robot's joint-link segment for a critically damped operation which is independent of the variation of moment of inertia due to the configuration of the robot kinematic geometry.

Thus it would appear that with a fixed but well-defined nominal value of the position feedback gain nk_p the overall loop gain can be set constant satisfying the condition of inequality in Equation 7.23. A variable velocity feedback gain nk_v would be sufficient at the first approximation to control the variable effective inertias and to ensure critical damping for any load at any stretch of the robot arm segment. Thus, a variable velocity feedback gain coefficient nk_v can keep the damping under control to obtain a critically damped response when the inertia varies within its maximum and minimum limits. The time response of such a system to a unit step or impulse input is an exponential function. Time responses will look like those depicted in Figure 7.6 for critically damped and overdamped systems.

It is well known that a definite and useful relationship exists between the time and frequency responses of a second order linear system. For a higher order model of a servo-system a similar but more complex relationship can be derived. In this respect, the derivation presented is quite useful and adequate as a general guide, but care must be taken not to extend the results excessively.

When approximating a higher order system by a second order model,

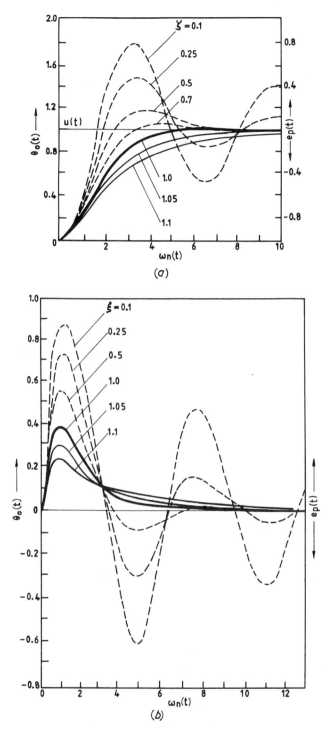

Figure 7.6 Time responses of a robot's joint-link arm segment. The heavy line is the response for $\zeta = 1$. (a) For unit step input $\theta_p(s) = \mathscr{L}[u(t)]$. (b) For unit impulse input $\theta_p(s) = \mathscr{L}[\delta(t)]$.

assume that any of the segments have a unit step and unit impulse response for $\zeta > 1$ approximately as depicted in Figure 7.6. When the system transfer function of a higher order system is dominated by a double real pole representing critical damping ($\zeta = 1$) or by a pair of real poles on the real axis representing overdamping ($\zeta > 1$), any of these pole pairs corresponds to an equivalent second order system. Although this type of approximation lacks theoretical justification, in practice it is quite acceptable in the design of robot servo performance because of the simplicity of the relationship between the time and frequency domain responses, which are valid for second order systems only. The development of a higher order model is particularly advisable at a later stage of the design. Then the exact mathematical relationship between the time and frequency domains is expressed by using either Laplace or Fourier transformations [7.7], [7.8].

7.4   ERROR RESPONSES AND STEADY-STATE ERRORS IN ROBOT SERVOS

In this section the simplified error response analysis of a robot joint-link segment servo-system will be described. In robot operation a sufficient knowledge of errors in general and of certain types in particular is of great importance.

Frequently there is more than one input and more than one type of input. In fact, we are dealing with multiple input and multiple output (MIMO) systems, remembering that in addition there are also various types of disturbance. First we consider the effect of each error separately, and then, if we need the complete error response, the principle of superposition can be applied. The resultant performance of all joint-link servo systems is judged on the ability of the individual controls to reproduce accurately the set inputs and to minimize the unacceptable effects of the disturbances wherever and whenever they may occur in the robot servo loops.

Primarily we are dealing with conventional servo-system errors, i.e. position, velocity and acceleration errors with a great emphasis on the derivation of their steady-state values. In addition there are dynamic errors caused by the disturbances in the robot system. Such disturbances are generated by random changes of load and/or gravity load; noise and drift in the amplifiers and digital circuits; coulomb friction and backlash in the motor, gear-train and transmission belts or chain drives, etc. To enhance the control performance, there is, in addition to the position input, a separate set point for velocity input to keep the speed under control for trajectory tracking purposes.

Regardless of the type of control system, the objective of the servo is to generate the required torque and/or force at the joints for accurate positioning of the output shaft at the target points at specified speeds. To be able to run the robot at higher speeds the controller's parameter changes may cause problems depending upon the arm's geometry configuration. So there are various detrimental effects which must be taken into consideration during the design of robot servos to counteract or at least to reduce the errors to a minimum level.

Now the main objective is to derive the position, velocity and acceleration error responses, and from these, wherever possible, to determine the steady-state errors. Whenever their reduction is desirable, compensation is required either to be introduced within the servo loop or to provide torque to the joints in a feedforward manner in order to eliminate the steady-state errors and at least partly to reduce the dynamic errors.

Position error response and its steady-state value

In the following we will examine the position errors due to

- (i) A position input command $\theta_p(s)$.
- (ii) A disturbing torque $\theta_T(s)$.
- (iii) The coulomb friction and stiction.
- (iv) The gravity loading.

In order to derive the various position error responses and their steady-state values we will modify the basic block diagram in Figure 7.7(a). We move the rate feedback loop summing point to behind the Ak_m block and eliminate the damping feedback loop by applying block algebra [7.3], [7.4]. The modified block diagram of the system is shown in Figure 7.7(b).

(i) The position error response $e_p(s)$ due to the set position command input $\theta_p(s)$ ($\equiv q_i(s)$, consult Chapters 5 and 6) is derived from the block diagram of Figure 7.7(b) as

$$e_p(s) = G(s)\theta_p(s) \tag{7.26}$$

where

$$G(s) = \frac{e_p(s)}{\theta_p(s)} = \frac{s^2 n^2 J'' + (D + Ak_m nk_v)s}{s^2 n^2 J'' + (D + Ak_m nk_v)s + Ak_m nk_p} \tag{7.27}$$

The steady-state position error $e_{p,ss}$ due to a constant position step input $\theta_p(s) = \theta_p/s$ is from the final value theorem given by

$$e_{p,ss} = \lim_{s \to 0} [se_p(s)] = 0 \tag{7.28}$$

The steady-state position error for a Type-1 system is always zero; a constant

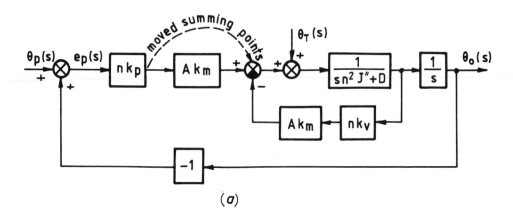

(a)

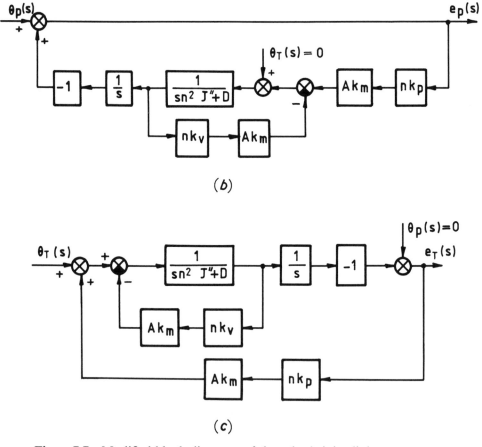

(b)

(c)

Figure 7.7 Modified block diagrams of the robot's joint-link segment servo-system for: (a) Block diagram of Figure 7.4(a). (b) Position error response $e_p(s)$ due to the set command $\theta_p(s)$. (c) Error response $e_T(s)$ due to the torque disturbance $\theta_T(s)$.

actuating error produces a constant rate of change of the controlled variable. In other words with zero actuating position error, the controlled variable must be equal to the reference input since there is no position error.

(ii) To demonstrate the effect of a disturbance in a robot joint-link segment servo-system, we will introduce a disturbing torque $\theta_T(s)$ at the summing point just beyond the Ak_m block (see Figure 7.7(a)).

The position error response $e_T(s)$ (rad), due to the torque disturbance input $\theta_T(s)$ (N m) can be derived from the block diagram in Figure 7.7(c), and is given by

$$e_T(s) = G_T(s)\theta_T(s) \tag{7.29}$$

where

$$G_T(s) = \frac{e_T(s)}{\theta_T(s)} = \frac{1}{s^2 n^2 J'' + (D + Ak_m nk_v)s + Ak_m nk_p} \tag{7.30}$$

In this case the steady-state error due to a constant torque step input $\theta_T(s) = \theta_T/s$ is given by

$$e_{T,ss} = \lim_{s \to 0} [se_T(s)] = \text{constant} \tag{7.31}$$

$$e_{T,ss} = \lim_{s \to 0} \left[s \frac{1}{s^2 n^2 J'' + (D + Ak_m nk_v)s + Ak_m nk_p} \frac{\theta_T(s)}{s} \right] \tag{7.32}$$

$$e_{T,ss} = \frac{\theta_T}{Ak_m nk_p} \tag{7.33}$$

The steady-state error arising from a constant torque disturbance is known as the offset from which the servo stiffness is obtained in (Nm/rad) as

$$\frac{\theta_T}{e_{T,ss}} = Ak_m nk_p \tag{7.34}$$

(iii) The coulomb friction and stiction of the system must be overcome before the motor-gear unit will start to move. Once this unit is in motion, the viscous friction takes over and opposes the motion proportional to velocity. Stiction error must be carefully considered so that it can be related to repeatability, which is one of the most important performance figures in the robot's specification. The coulomb friction error $e_c(s)$ is usually compensated for by introducing an appropriate torque or force, applied to the joints, via a feedforward loop.

(iv) The gravity loading may or may not have an effect on a particular joint-link manipulator segment. In any case we need to know all the effective loads and the values of the effective inertias either by calculation or measurement. It is easy to establish that the shoulder and the elbow are the joints most sensitive to gravity loading. These joints therefore require more attention in the design of their servo systems. The rest of the joints have less limiting constraints and their gravity loading does not usually need to be taken into consideration for compensation. But for the two main joints, namely the shoulder and elbow, we should apply an additional feedforward torque to the motor servo to balance out the error caused by gravity (see Section 7.5). The block diagram of the joint-link servo-system for various combinations of position error and torque compensation, as discussed above is shown in Figure 7.11.

Velocity error response and its steady-state value

Velocity error may cause concern when the robot operates in coordination with a moving system, for instance picking up pallets from a moving conveyor belt, or if the technology accomplished by the robot requires a fixed speed, e.g. arc welding or paint spraying.

Velocity errors are caused by the following effects:

(i) The set velocity input command v_c.
(ii) Coriolis and centripetal forces, when these are not negligible.
(iii) The variation in effective moments of inertia during robot operation.
(iv) The moments of inertia of the joint-link segments may interact by mutual coupling which could cause oscillation in the case of nonlinearities.

(i) To obtain the velocity error response $e_v(s)$ we apply a constant velocity command input v_c at the set point ($\equiv \dot{q}_i(s)$, consult Chapters 5 and 6). Let us designate the Laplace transform of a constant velocity v_c by $v_c(s)$, then the velocity error response is expressed by

$$e_v(s) = G(s)v_c(s) \tag{7.35}$$

The steady-state velocity error due to a constant step velocity input $v_c(s) = v_c/s^2$ is, by the final value theorem, given by

$$e_{v,ss} = \lim_{s \to 0} [se_v(s)]$$

By substituting for $G(s)$ from Equation 7.27 into Equation 7.35, $e_{v,ss}$ becomes

$$e_{v,ss} = \lim_{s \to 0} \left[s \frac{s^2 n^2 J'' + (D + Ak_m nk_v)s}{s^2 n^2 J'' + (D + Ak_m nk_v)s + Ak_m nk_p \, s^2} \frac{v_c}{s^2} \right] \tag{7.36}$$

$$e_{v,ss} = \frac{D + Ak_m nk_v}{Ak_m nk_p} v_c \tag{7.37}$$

$$e_{v,ss} = \frac{D}{nk_p Ak_m} v_c + \frac{k_v}{k_p} v_c \tag{7.38}$$

Assuming critical damping and applying Equations 7.19 and 7.16 the steady-state velocity error, usually referred to as the tracking error, becomes

$$e_{v,ss} = \frac{2\sqrt{Ak_m nk_p n^2 J''}}{Ak_m nk_p} v_c \tag{7.39}$$

$$= 2\sqrt{\frac{n^2 J''}{Ak_m nk_p}} v_c$$

A typical value of the tracking error can easily be computed using for instance a midrange structural frequency

$$e_{v,ss} = \frac{2}{w_{st,mid}} v_c = \frac{1}{\pi f_{st,mid}} v_c \tag{7.40}$$

This steady-state velocity error along the motion trajectories usually causes detrimental tracking errors. Large tracking errors would be intolerable in painting or welding robots. Their complete elimination, or at least partial reduction, is of paramount importance in speed controlled servo systems. This may be achieved by providing both feedback and feedforward torques, one part of which is used to overcome the damping effect of D in the subsidiary feedback loop, and the other to keep under control the effect of the introduced damping nk_v in the negative feedback loop. In fact for the compensation of the steady-state velocity error, a Type-2 system would be required [7.3]. The joint-link servo-system with velocity compensation is shown in Figure 7.8, which is derived from Equation 7.38.

(ii) Coriolis and centripetal forces only become significant when the robot operates at high speed. Otherwise their effects are minimal. Fortunately high positional precision and high speed are seldom of equal importance. For instance in high precision machining, near to the target position the robot is

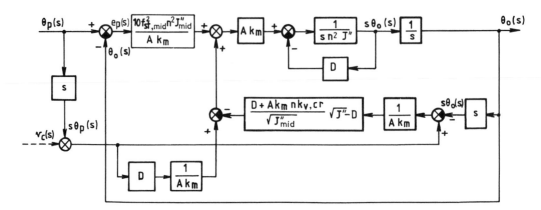

Figure 7.8 Block diagram of steady-state velocity error compensated servo of a joint-link arm segment.

expected to move slowly to keep high positioning accuracy, while in the course of moving from one target to the next one in free flight, high speed is more important than accuracy, i.e. accurate path tracking. We may note that Coriolis and centripetal forces can be derived by the use of Lagrangian mechanics by solving the manipulator's dynamic equations. The interested reader is referred to the literature [7.5].

(iii) and (iv) To deal with the effect of inertias and mutual coupling would require a knowledge of the effective distributed mass and moment of reflected inertias, the coupling effects between the subsequent joint-link arm segments and the effective forces and torques, etc. The derivation of these parameters is possible, for instance, by Lagrangian calculus, but this is beyond the scope of basic robotics.

Acceleration error response and its steady-state value

The acceleration error response $e_a(s)$, due to a constant acceleration command input $a_c(s)(\equiv \ddot{q}_i(s)$, consult Chapters 5 and 6) either by setting its value or by accelerating the position input θ_p, is given by

$$e_a(s) = G(s)a_c(s) \qquad (7.41)$$

The steady-state acceleration error is obtainable from the acceleration error response $e_a(s)$ by employing the final value theorem

$$e_{a,ss} = \lim_{s \to 0} [se_a(s)]$$

As the Laplace transform of a constant acceleration command input $\mathcal{L}[a_c(s)] = a_c/s^3$ is introduced, the steady-state acceleration error is obtained as

$$e_{a,ss} = \lim_{s \to 0} \left[s \frac{s^2 n^2 J'' + Ds + Ak_m nk_v s}{s^2 n^2 J'' + (D + Ak_m nk_v)s + Ak_m nk_p} \frac{a_c}{s^3} \right] \tag{7.42}$$

The final value of the $e_{a,ss}$ due to the first term in the numerator means that the system has a constant acceleration error, namely

$$e_{a,ss} = \frac{n^2 J''}{Ak_m nk_p} a_c \tag{7.43a}$$

The second term is zero if the velocity error has already been compensated. The third term would cause very large steady-state acceleration error which may be reduced to some extent via feedforward compensation.

A typical value of the steady-state acceleration error can readily be computed by using, for instance, the midrange structural frequency at a given speed

$$e_{a,ss} = \frac{1}{w_{st,mid}^2} a_c = \frac{1}{4\pi f_{st,mid}^2} a_c \tag{7.43b}$$

Assuming that the velocity feedforward signal is already employed to eliminate all velocity dependent errors, the second term in the numerator does not provide steady-state acceleration error. If velocity compensation should not have been employed, this acceleration error would approach infinity with time, which would mean that a Type-2 system cannot follow a parabolic input. Its compensation needs the introduction of a second integrator in the forward path of the servo-system. In conclusion, for full compensation of the steady-state acceleration error, a Type-3 system would be required [7.3].

The acceleration error is mostly encountered at the start and the end of the motion. For more accurate trajectory tracking it is advisable to introduce compensation for all acceleration errors by adding another feedforward term $s^2 n^2 J''$. If the effective inertias of the joints are not known, as is usually the case, then the minimum inertia J_{min} can be used, bearing in mind that only partial compensation for the acceleration error is being applied. The use of the maximum inertia J_{max} would result in an underdamped response having some overshooting motion when the joint inertia was less than J_{max}. The joint-link servo-system, with acceleration error compensation, is shown in Figure 7.9.

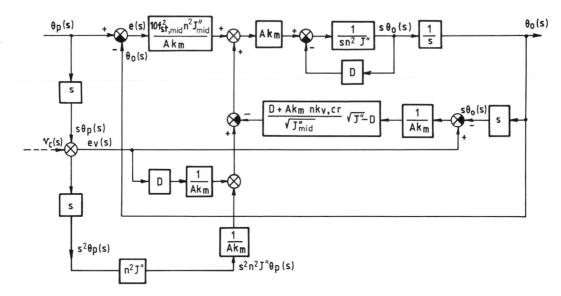

Figure 7.9   Block diagram of steady-state acceleration error compensated servo of a joint-link arm segment.

7.5   FEEDBACK AND FEEDFORWARD COMPENSATIONS

In Section 7.4 we carried out an error response analysis and developed the motion models which consisted of position, velocity and acceleration responses. The performance design of robots is centered around their position and velocity characteristics with particular emphasis on the types of compensation.

As far as robot technology is concerned, it may occur that accuracy of both position and speed must be achieved. These are two conflicting requirements and a compromise between them determines to what extent effects can be neglected during the design when simplifications are considered. In this area a great deal of uncertainty exists; the design may rely either on position or velocity but not simultaneously on both. We may call this the 'uncertainty principle' of robot design. In essence, this principle states that it is impossible to achieve simultaneously, both the exact position and the exact speed at any one instant of time. Thus, instead of speaking about accurate, simultaneous values of these two variables, we may say that at certain speeds there will be a probable position accuracy or vice versa. This is the reason why robot design in general, and compensation in particular, occurs with respect to either position or speed accuracy but not to both. Whichever the design objective, the best starting point at this stage is to define the speed specification, regardless of the tracking accuracy requirement. The usual speed characteristics are depicted in

Figure 1.15(a). These are linear functions of triangular or trapezoidal shape, having a motion period T and a maximum velocity v_{max} at $T/2$.

In Section 7.4 we dealt with the error response analysis assuming that all elements operate in a linear fashion. But in robot systems this is far from the truth. Many forms of nonlinearity are present, of which stiction and coulomb friction are illustrated in Figure 7.10.

In the following we will deal with compensation in an interpretive manner without going into the mathematical details. To compensate for the torque error due to coulomb friction, we introduce feedforward torques to overcome

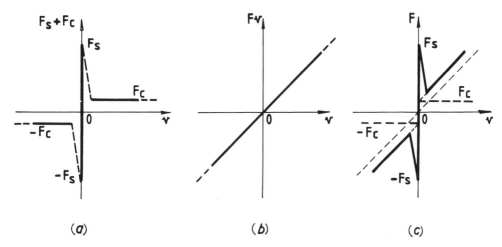

(a)   (b)   (c)

Figure 7.10 Nonlinear and linear forces versus velocity in robotic servos. (a) Stiction force F_s and coulomb friction force F_c. (b) Viscous friction force F_v. (c) Combined linear and nonlinear forces $F = F_s + F_c + F_v$.

both the static and dynamic friction torques, $T_{st}(s)$ and $T_d(s)$ respectively. The blocks representing this torque error compensation are shown in Figure 7.11. Stiction and all the other higher order nonlinearities such as dead zone, backlash with hysteresis and limiting effects are not dealt with. These considerations are beyond the scope of this book, but you may find excellent treatment of these problems in the literature [7.9].

The steady-state error due to gravity loading may cause problems under maximum load particularly with the arm in full stretch. This also depends upon the robot's kinematic geometry. But when the load is excessive, the compensation for gravity loading should provide an additional feedforward torque $T_g(s)$ at the appropriate summing point in the forward path [7.10] as shown in Figure 7.11.

We may consider a partial compensation for the steady-state acceleration error usually at the beginning and end of the motion. It is possible to

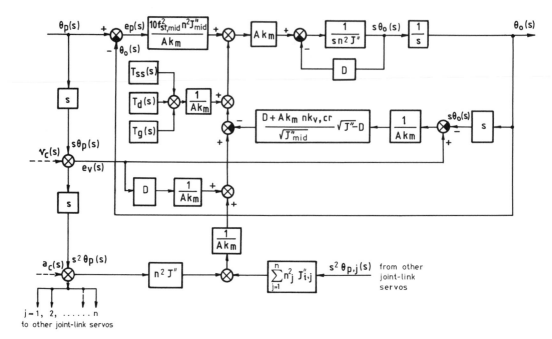

Figure 7.11 Coulomb friction, gravity loading and joint coupling compensations, including the steady-state position, velocity and acceleration error compensations.

compensate for these steady-state errors by adding the following feedforward term (see Figure 7.11):

$$T_a(s) = n^2 J'' s^2 \theta_p(s)$$

In the high speed region of the curve there still remains a small velocity variation, but for that acceleration no compensation is necessary.

Finally when we free all the joints, the inertial coupling, centripetal and Coriolis forces will be free to act and interact; so there will be further problems in eliminating those effects. Without going into details, we still indicate the intercoupling effects between the joint-link arm segments. Of course any single joint-link affects all other servos as shown by the acceleration output links in Figure 7.11.

7.6   QUALITATIVE FEATURES OF ROBOT SERVOS

In this section before we consider a computer controlled servo-system analysis for error compensation, it is useful to develop qualitatively some principal ideas concerning the performance of servo-systems in terms of control actions, i.e. proportional, proportional+derivative and proportional+integral control [7.11]. In the system shown in Figure 7.3, an electric motor is used for positioning a load consisting largely of pure inertias. The output and input position angles θ_o and θ_p are compared and the difference between them produces a proportional error signal e_p. In addition, there are signals corresponding to the derivative and/or integral of the proportional position error, with respect to time, which operate superimposed on the proportional error signal.

Proportional (P) control

In a proportional servo-system the motor is assumed to develop a torque substantially proportional to the total error applied to the input of the amplifier. If the input is suddenly rotated through an angle θ_p, the error will cause a torque on the load which will rotate it in the appropriate direction to reduce the error. Proportional control system responses of critical settings are shown in Figure 7.12(a).

Since the error does not become zero until the system aligns, there will be a continuous, although decreasing, torque in the same direction applied to the load. The load velocity will increase continuously to reach its maximum value. When the system runs through the alignment position the error becomes zero. As a result of its inertia the load overshoots the set value, the error and hence the torque now reverse direction, the system slows down and comes momentarily to rest beyond the required position at the top of the overshoot. The reversed torque then accelerates the load back towards alignment, but the system will undershoot below the set value (see the dotted line in Figure 7.12(a)). The oscillation may build up, remain constant or die away. In robot servos none of these are allowed to occur.

In servos utilizing substantially proportional control, the essential problem is that there is no reversal control torque before the system aligns, so that it is bound to overshoot badly. For this obvious reason, in robot control we should not allow even the slightest oscillatory operation and we have to introduce various interoceptive feedback and feedforward compensating actions outlined in Section 7.5.

Proportional+derivative (PD) control

In Figure 7.12(b) the error for a critically damped system response is plotted separately and its slope is the derivative of error. Since the error is decreasing the derivative error is negative and would reach its maximum value when the error is passing through its inflexion point. If the proportional error and its

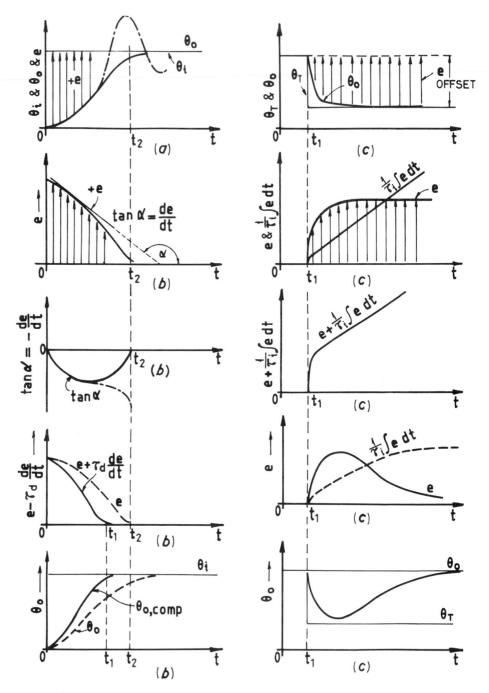

Figure 7.12 Compensation types in critically damped robot servo-systems. (*a*) Proportional control (PC). (*b*) Proportional+derivative control (PDC). (*c*) Proportional+integral control (PIC).

derivative are added together, their sum will be zero before the actual servo-system is aligned. So the system is controlled by

$$(e_p + \tau_d \, de_p/dt).$$

The output torque will reverse before output alignment occurs and effectively acts as a brake which greatly reduces the transient. Thus in qualitative terms, derivative control can be applied to control the slope of the position response, i.e. the response time will be shorter, $t_1 < t_2$, as shown in Figure 7.12(*b*).

Proportional + integral (PI) control

Suppose that a constant disturbing torque is externally applied to the load reflected to the robot joint. As this torque rotates the load, an error signal will be produced which will generate an opposing torque. After the transient the system will become steady, but will be misaligned by a resulting steady-state error. If the disturbing error was substantially a step input, by integrating it there will be ramp error

$$1/\tau_i \int e_p \, dt$$

and the system will be controlled by the proportional error plus the integral of the error, i.e.

$$e + 1/\tau_i \int e_p \, dt.$$

Thus after the load disturbance has misaligned the system, the integrator output will increase and the control torque generated will exceed the disturbing torque, so that the surplus torque will overcome the steady-state misalignment. As the load swings back towards the position of alignment, the error decreases. In the steady-state operation the error is zero, but the output of the integrator remains constant at a certain value, which gives adequate torque to maintain alignment of the system (see Figure 7.12(*c*)). In conclusion the purpose of integral control is to eliminate the steady-state error known as offset.

These three types of compensation illustrate some of the problems encountered in robot servo-systems and indicate what additional actions may be required in both forward and feedback paths in order to achieve a satisfactory operation. The question briefly is how do we introduce the so-called PID control actions. In the case of robot servos those control actions can be applied in numerical terms in the microprocessor program in order to derive the software option of the derivative and integral compensation.

7.7 COMPUTER CONTROLLED SERVO-SYSTEM FOR ROBOT APPLICATIONS

It is essential to have the facility of microcomputers in the servo-systems controlling the movements of the joint-link arm segments of the robot so that good flexibility of the multipurpose, multifunction robot arm can be achieved.

The computer controlling the robot stores in its memory the sequence of movements and operations that the robot arm requires to perform in order to achieve a particular objective. This may be as simple as picking up an item at point A and moving it to point B. On the other hand it may be as complex as picking up an object, with correct orientation, from a moving conveyor belt. Whatever the control objectives of the robot are, a well-designed microcontroller and an intelligent controlling servo algorithm are required.

The computers control and monitor the performance of the robot arm as well as interact with other sensing and acting equipment in the environment of the robot. Interaction requirements for a robot may take one of many forms. Here we concentrate on two schemes that have already been indicated in Section 7.1.

We consider first a single processor system, fast and powerful enough to handle the control of all joints and at the same time able to monitor all housekeeping work as the master computer of the robot system. Alternatively a multiprocessing scheme based on medium power processors may be used, one of which is the master and the rest are the so-called comicros each controlling a single arm joint only.

For a single microcontroller system nowadays, a good choice could be based on the currently available Motorola 68000 16/32-bit processor or on one of the new generation of true 32-bit processors. For a multiprocessor system we would suggest a combination of the Intel 8088 16-bit processor as the master and the Zilog Z80 or Rockwell 6502 8-bit processor as the coprocessor for the joint controllers. The choice of the most suitable computer system depends on the complexity of the structural geometry of the robot and the kinematic performance required for its intended application, i.e. on the complex nature of the robot assisted process.

For small size and low speed robots a single 8-bit processor would be adequate. For medium size robots of medium speed and high resolution, a multiprocessing system with 16/32-bit master processor would be appropriate for reasons of performance and economy. For large robots, when high speed is required and consequently low accuracy can be tolerated, any single 8-bit processor would be adequate. For high performance multifunction assembly robots a multiprocessing controller system would be preferable. For instance, the Unimation PUMA 600 employs a DEC LSI-11/23 16-bit micro for the master and a Rockwell 6503 processor for each of the axes controlled.

In robot control a distinction may be made between two categories. The first category, described as the low level control, is for controlling the position and speed of the joint axes using the information from the interoceptive sensing devices. This may involve the use of potentiometers, incremental or absolute shaft encoders attached directly or geared to the arm joint motor shaft.

The second category may be described as high level control which defines the functional motion. The computer obtains the data from both interoceptive and exteroceptive sensors, modifies data according to the required kinematics and returns the modified data to the joint servos in order to control the position and speed, and to monitor the functional requirements for motion.

The controlling computer will hold in its memory the information (and this may be continuously updated as the robot learns about its environment and itself) and algorithms that define the characteristics of the robot arm to compute the kind of movements that the arm will make between any initial and target points.

In order to move the tool center point (TCP), for example, along a straight line between the two points A and B, the master computer must have knowledge of the kinematic model of the arm, and activate the motor of each joint independently in such a way that the end effector would move from point A to point B at the predefined speed. As the robot moves on the straight line AB, all motors will rotate at different speeds for different periods of time and some may not rotate at all, so that the resultant movement of the TCP at the end of the robot will be on a straight line between A and B.

The servo controllers have no knowledge of the kinematic model of motion they perform. Their only function is to drive the motors to the required positions and to maintain the desired speeds for a defined period, as set by the master computer. On the other hand the functional motion controller, that is the master, knows the kinematic model in great detail and the type of motion to be performed. It calculates the position and speed for all joint axes of the robot in the selected coordinate frame and continuously provides that information to the servo controllers.

The use of a multiprocessing scheme in a robot system is preferable and can be implemented with no particular difficulties. It is electronically a modular structure and consequently a more reliable and serviceable system because the control of each axis is individual, so a single microprocessor handles the control for only one axis. All the processors continuously and simultaneously take position and speed data from the output shafts via analog-to-digital converters and at the same time they report the status of their arm joints (speed, position and orientation) to the master computer. The master compares these data with the set point values and computes the motion function. The master computer is now able to calculate the proportional, integral and derivative signals for PID compensation and provides the desired data for each axis

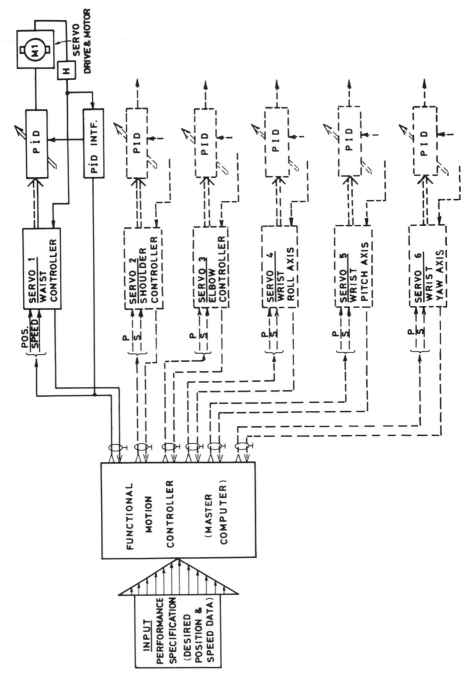

Figure 7.13 Computer controlled servo-system with functional motion control loop.

controller. These are required to be fed, via a suitable interface, i.e. digital-to-analog converter, into the arm joint servo-systems in order to provide the necessary corrective actions for axis misalignments during the motion (see Figure 7.13). In fact there will be a critically damped time response with improved response time.

The functional motion controller (the master) also performs the inverse coordinate transformation which means that the time required for computing the joint coordinates should be considerably less than the response time of the individual joint servo-systems. Advanced second generation robots usually have three modes of coordinate operation, i.e. in joint, world and tool coordinate frames.

We may mention from experience that, in the case of fast motion, the integral compensation for arm-joint servos is of little significance. The reason for this is that the robot moves around for most of the time and steady-state operation occurs only rarely. Integral control needs time to correct the position (steady-state error) and there is often not sufficient time before the robot is required to move on from the target points. If we try to use a large integral constant to make the integration faster acting, we will find that the system, due to the excessive integration, becomes unstable. The question now is what other method can we use for the reduction of the steady-state error instead of integral compensation.

In order to reduce the steady-state error to a minimum value in the output response, to improve the accuracy and repeatability and at the same time to maintain stability at a favorable level in all servos, we propose a novel fast-acting compensation without integration. It can be shown that in a PD controlled system taking a few samples, a minimum of three, at the initial part of the time response, is quite sufficient to estimate the response slope. Knowing also the set value of the response stated in the program, it is possible to obtain the functional control of the time constant by setting it to give a high slope for a period of about 90 percent of the response time and then to finish off with a critically damped response just before the overshoot would start. This kind of composite compensation, by applying a fast and then a slowed down derivative action, will result in an enhanced critically damped operation. But its application requires the introduction of a parameter estimator with a reasonably fast sampling facility and a simple algorithm to compute the time constant.

The condition of critically damped compensation has been shown in Section 7.3. A simpler version of two-term control has been derived which provides only the limiting values of the relevant parameters for a two-term feedback controller to safeguard a critically damped operation regardless of load and stretch of the robot arm.

A sophisticated three-term PID controller can also be designed with variable P, D and I in an adaptive manner to set the system in an optimum operating condition. But for this purpose a highly sophisticated parameter estimator should be involved and a fast operating algorithm computing the parameter values is required in order to maintain the system always in critically damped operation and to reduce steady-state errors. The involvement of a parameter estimator and its control by algorithm readily indicates the complexity of three-term PID control in robot servo applications.

From a robot stability point of view it is worth mentioning that the analog-to-digital and digital-to-analog converters may create problems as part of the forward path and feedback loop. The sampling and conversion time may present a lag in the system which may be too long, therefore expensive high speed converters have to be used. The resolution of the converters may be a limiting factor in the achievement of the overall system performance specification. Normally there has to be a trade-off between conversion speed and resolution with appropriate compensation built into the computational model used to derive the control outputs.

With additional adaptive servo intelligence it would be possible for the computer to improve further the performance of the robot system and to change dynamically the system model by controlling both feedforward and feedback signals during the program execution.

7.8   BENEFITS OF USING COMPUTERS IN ROBOT CONTROL

In practice computer based control is essential to achieve satisfactory robot performance since the tasks are defined by programs written by the user. The main benefits of using computers for controlling robots are:

(i) Flexible use of robot by interactively changing programs which allows modification of the manufacturing process or future product development and expansion.

(ii) Trajectory generation for improved performance and the availability of continuous trajectory computation permits complex motions.

(iii) The ability to work in various frames of reference such as polar, cylindrical revolute coordinate systems.

(iv) The ability to work in moving frames of reference.

(v) The ability to counteract unpredictable situations.

(vi) The ability to use interoceptive and exteroceptive information from sensors such as position, velocity, as well as touch, vision, voice, etc.

(vii) The ability to interact with a human operator permitting real-time modification and program generation.

(viii) The ability to provide easy access to pretested and predefined manipulation routines through the use of libraries.

7.9   APPENDIX

Appendix 7.A   Syke 600–5 robot; motor and gear-train assembly

All the three major robot joints consist of a d.c. servomotor, optical encoder, potentiometer and release brake. The d.c. servomotors for waist, shoulder and elbow are of samarium cobalt (rare earth) types. For the wrist pitch and wrist roll axes, permanent magnet d.c. servomotors are used. There are release brakes on each major axis and 24 V d.c. 0.45 A activate to release the brake. Individual joint assemblies are illustrated in exploded forms in Figures 7.A.1 and 7.A.2.

All motors, brakes, encoders and potentiometers are internally wired to two fixed connectors. One connector carries motor and brake drive signals, the

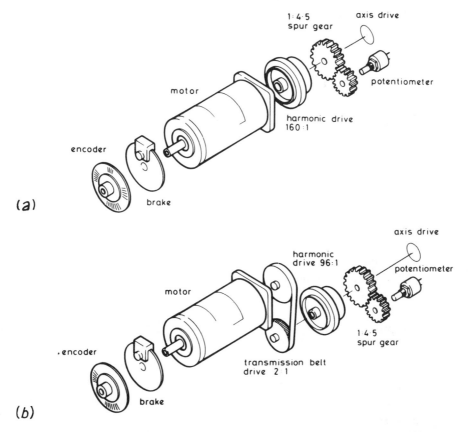

Figure 7.A.1   Robot major arm joints in exploded form. (*a*) Waist and shoulder drive. (*b*) Elbow drive.

other is for encoder and potentiometer signals. Specification of all mechanical components are listed as follows:

Waist, shoulder and elbow d.c. servomotors

Rated power	100 W
Max. operating speed	6000 rpm
Continuous torque (stall)	0.45 N m
Peak torque	2.15 N m
Current at continuous torque	3.65 A
Current at peak torque	18.00 A
Max terminal voltage	90 V
Torque sensitivity	0.122 N m A^{-1} + 10%
d.c. resistance	3.3 ohms
Inductance	3.0 mH + 30%

Wrist pitch and wrist roll d.c. servomotors

Rated speed	3000 rpm
No load speed	5000 rpm
Rated torque	0.08 N m
Peak torque	0.35 N m
No load current (max)	0.5 A
Voltage	12 V
Torque sensitivity	0.019 N mA^{-1}
d.c. resistance	0.75 ohms
Inductance	0.8 mH

Optical encoders (all axes)

Supply voltage	+5 V d.c.
Supply current	120 mA
Resolution	200 ppr (pulses/rev)
Max. sink current	4.0 mA
Max. source current	0.5 mA
Output device	L339 open collector (3K6 pull-up resistor)
Output	2 channel square wave with index pulse

Potentiometers (all axes)

Resistance	10 K ohms
Linearity	±0.2 %
Wattage	3 W
Turns	10

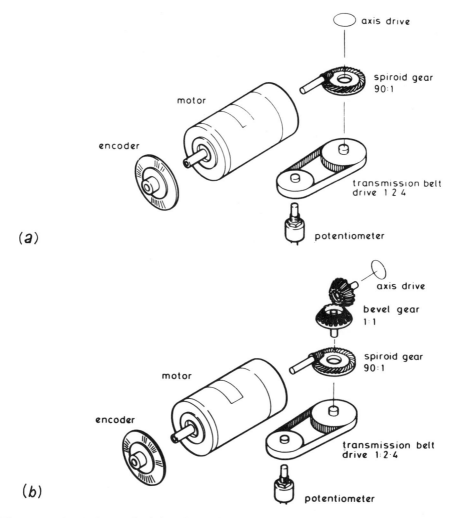

Figure 7.A.2 Robot wrist joints in exploded form. (*a*) Wrist pitch. (*b*) Wrist roll.

7.10   REFERENCES

[7.1]   Foster, D. *Automation in Practice*, McGraw-Hill, 1968.

[7.2]   Power, H. M. and Simpson, R. J. *Introduction to Dynamics and Control*, McGraw-Hill, 1978.

[7.3]   D'Azzo, J. J. and Houpis, C. H., *Feedback Control System Analysis and Synthesis*, 2nd edn, McGraw-Hill, 1966.

[7.4]   Kuo, B. C. *Automatic Control Systems*, Prentice-Hall, 1975.

[7.5]   Paul, R. P. *Robot Manipulators: Mathematics, Programming and Control*, The MIT Press, 1983.

[7.6]   Scheinman, V. D. *Design of a Computer Manipulator*, Stanford Artificial Intelligence Laboratory, Stanford University, *AIM 92*, 1969.

[7.7]   Stroud, K. A. *Laplace Transformation: Programs and Problems*, Stanley Thornes, 1978.
[7.8]   Chirlion, P. M. *Signals, Systems and the Computer*, Intext Educational Publishers, 1973.
[7.9]   Servant, C. J. *Control System Design*, Ch. 8, McGraw-Hill, 1958.
[7.10]   Bejzy, A. K. Robot Arm Dynamics and Control, *NASA-JPL Technical Memorandum*, February 1974, 33–669.
[7.11]   Snyder, W. E. *Industrial Robots: Computer Interfacing and Control*, Prentice-Hall, 1985.

INDEX